# POLITICS
### AND
# THE LAND

# POLITICS

## AND

# THE LAND

By

### CECIL DAMPIER-WHETHAM
#### M.A., F.R.S.
FELLOW AND SOMETIME SENIOR TUTOR
OF TRINITY COLLEGE
CAMBRIDGE

## CAMBRIDGE
### AT THE UNIVERSITY PRESS
#### 1927

CAMBRIDGE UNIVERSITY PRESS
Cambridge, New York, Melbourne, Madrid, Cape Town,
Singapore, São Paulo, Delhi, Mexico City

Cambridge University Press
The Edinburgh Building, Cambridge CB2 8RU, UK

Published in the United States of America by Cambridge University Press, New York

www.cambridge.org
Information on this title: www.cambridge.org/9781107615526

© Cambridge University Press 1927

First published 1927
First paperback edition 2013

*A catalogue record for this publication is available from the British Library*

ISBN 978-1-107-61552-6 Paperback

# PREFACE

Perhaps an excuse is needed for the publication of yet another book on agricultural policy, or, at all events, an explanation of how it comes to be written.

The author is a landowner on a small scale, both by purchase and by inheritance, and farmed part of his land for eight years. He is also a Fellow of a Cambridge College, who, in the intervals of teaching and of University and College administration, has made some study both of economics and of agricultural science. Of late years as a member of the Lawes Agricultural Trust and of the Council of the Royal Agricultural Society, he has had the advantage of association with many leaders in both the theory and the practice of agriculture, but of course he alone is responsible for the contents of this book.

The present work is founded on articles dealing with the Economics of Agriculture published in the *Journal of the Royal Agricultural Society*, vol. LXXXV, 1924, and in the *Economic Journal* for December 1925 and March 1926. In view of the public interest now taken in the past and future of English land, and of the prospect of our system of land tenure becoming an active political question, the author was asked to expand those articles and adapt them to a wider circle of readers. He wishes to thank all those who have helped in the preparation of the book—in particular the landowners, bursars of Colleges and land agents who have put their rent books at his disposal; Mr J. A. Venn for reading the proof-

sheets, and Miss Christine Elliott for secretarial assistance.

Many books, reports and articles have appeared within
the last few years dealing with the state of agriculture
and the future of the land. It may perhaps be found
useful to have the chief results of recent enquiry brought
together in a convenient form, with a critical account
of the many proposals which have been made for rural
reconstruction and development.

In spite of all the discussion which has taken place,
there still seems to be no general understanding of the
economic conditions which determine the methods of
British farming, of the causes which produce agricultural
prosperity or adversity, or of their bearing on agricultural politics.

For his own satisfaction, the author has spent much
time and trouble in an attempt to understand these
conditions and causes. In the hope that the facts and
inferences which have helped him may be useful to
others, he has set them forth in the following pages.

CAMBRIDGE
20th December 1926

# THE LAND

*They talk about the land—the land*
    *So patient and so old,*
*So old when man first scarred it and*
    *So young when all is told:*

*The little fields that airmen see,*
    *Faint patch-work on the downs,*
*The long-forgotten husbandry*
    *That fed the British towns;*

*And fading lines upon the chalk,*
    *Look! trace them even now—*
*The open strip, the grassy baulk,*
    *Cut by the Saxon plough;*

*West-country bank and winding lane,*
    *From waste and wood compact,*
*Square Midland fields that tell so plain*
    *Of an Enclosure Act;*

*The running brook, the Doomsday mill,*
    *The Church tower firm and grey,*
*The Tudor cottages that still*
    *Are children's homes to-day;*

*The land our Northern sires laid out*
    *In Hams and Ings and Tuns,*
*That bred the Crecy archers stout*
    *And manned Lord Nelson's guns;*

*The land of those of vision true,*
    *With knowledge quick to arm,*
*Who from the old drew forth the new*
    *And taught the world to farm;*

*The land we love with hearts aflame,*
    *High hearts in joy or sorrow,*
*Through every change is still the same—*
    *So fear not change to-morrow!*

C. D.-W.

# CONTENTS

## PART I

### BRITISH AGRICULTURE

## PART II

### THE LAND AND ITS OWNERS

# PART III

## *THE FUTURE OF RURAL ENGLAND*

# POLITICS AND THE LAND

## INTRODUCTION

D uring times of agricultural depression it invariably happens that there is copious discussion of causes and proposal of remedies, (1) by men who get their living from the land, (2) by others who study economics, and (3) by some who suffer from neither of these obstacles to facile generalization.

In former times, agriculturalists seldom appreciated the underlying economic causes which make for prosperity or adversity, while few economists had an adequate knowledge of the details of practical farming. But, in the depression of the last five years, the modern schools of agricultural economics have brought their knowledge to bear and helped to prepare a series of Reports by Commissions, Tribunals and Committees, which have put at our disposal a wealth of old and new facts and many valuable conclusions. Among these Reports should be mentioned as specially helpful, those of Lord Linlithgow's Committee[1], of the Agricultural Tribunal of Economists[2], and of the Committee of the Ministry of Agriculture on the Stabilization of Prices[3].

The depression has not been confined to this country, and it is well to trace the similar course of events in the United States of America as set forth in the *Report* of the Joint Commission of Agricultural Inquiry[4].

[1] *Departmental Committee on Distribution and Prices of Agricultural Produce, Final Report*, 1924, Cmd. 2008.

[2] *Agricultural Tribunal of Investigation, Final Report*, 1924, Cmd. 2145.

[3] Ministry of Agriculture Economic Series, No. 2, 1925.

[4] *The Agricultural Crisis and its Causes*, Washington, 1921.

Of many books which have appeared since the depression began, special reference should be made to Mr J. A. Venn's *Foundations of Agricultural Economics*[1], to Mr R. R. Enfield's *The Agricultural Crisis 1920–1923*[2], to Lord Ernle's *The Land and its People*[3], and to an article by Mr C. S. Orwin on *Commodity Prices and Agricultural Policy*[4]. A study of these authorities will enable the enquirer to obtain a clear idea about the past and present state of the agricultural industry, both as a craft and as a business, and to form an independent opinion as to possibilities of future improvement and methods of development.

Before remedies are considered, it is well to study the nature and causes of the disease. About its nature there is no difference in responsible opinion. What agriculture suffers from is the existence of prices for agricultural produce which in general are unremunerative in view of the costs of production. Prices broke in 1920, fell rapidly till 1923, and since then have oscillated round a level about constant. Farmers of arable land suffered heavy losses during the first fall in values. Skilful men made some profit in 1924 and 1925, but the present prospect is not favourable.

The farmer is less able to bear a continued fall in prices than other producers, chiefly because of the length of time his operations take. His costs begin months before he sells his produce, and, when he goes to market, he may find that prices have fallen away till his receipts are less than the expenditure he has incurred at a higher

[1] Cambridge University Press, 1923.
[2] Longmans, 1924.
[3] Hutchinson and Co. 1925.
[4] *Journal of the Royal Agricultural Society*, vol. LXXXIII, 1922.

level. Because of the slowness of the turnover, the length of the economic lag, rising prices mean prosperity and falling prices adversity to the farmer more than to other men, and the movement of prices is to him more important.

For an adequate examination of the causes of change in agricultural values we must turn to the *Report* of the Committee on the Stabilization of Agricultural Prices, or to Mr Enfield's book on *The Agricultural Crisis*.

Plenty or scarcity, acting on ill-organized markets, causes a rise or fall in price of single crops—wheat, it may be, or potatoes. But the general change in agricultural prices, which causes universal prosperity or adversity, is always accompanied by a corresponding change in the prices of other commodities. The change is not confined to agriculture, and the cause is not agricultural. It can only be elucidated by the general economic theory of prices. When in a future chapter we deal with this question, we shall find that the chief factor involved is the amount of currency and credit available compared with the amount of business to be financed.

The long agricultural depression of 1875 to 1896 was due to the demand for gold and credit based on it outrunning the supply. Prices fell and farmers were distressed not only in England, but in heavily protected countries, and even in the new lands whose competition was and is almost universally blamed for the ills of the British farmer. It was only the discovery in 1886 of new South African goldfields, and the gradual increase in world currency and credit which followed, that caused an agricultural revival about 1900.

Similarly, the inflation in credit necessary for war

caused a rise in prices from 1914 to 1920, and, in turn, the rise in prices called for more currency and credit. The process became cumulative, cause and effect acting and reacting on each other. In the hectic boom of 1919, there almost seemed a danger that control would be lost, and the pound follow the rouble and the mark into nothingness. But a check was called, credit contracted, and in 1920 prices, agricultural and other, came tumbling down. Deflation followed inflation. The pound was saved, but industry, manufacturing and agricultural, became depressed, and unemployment grew apace.

When the general price level is falling, agricultural depression is inevitable. No skill in cultivation, no co-operative marketing, no import duties of possible amount, can prevent depression, though they may mitigate its effects. The only complete cure would be a stabilization of the general price level, for the chief cause of depression is not agricultural or economic, but monetary. This is the fundamental fact of the whole subject, and, till it and its consequences are understood and faced, it is useless to toy with partial remedies in the hope that they will cure general ills. Too often they are but treatment of the symptoms, which do nothing to check the progress of the disease.

I am not decrying the usefulness of some of them. Indeed it may be that alleviation of symptoms is all that is possible in present conditions, and will perhaps give time for Nature (in the form of a new goldfield) or Man (in the form of a refusal to produce enough food at present rates) to effect a radical cure. Yet I find it difficult to believe that we shall always consent to deal in terms of a unit of money the value of which depends

on the varying supply of, and demand for, a single commodity. What should we think of a standard of weight
or length which expanded by 40 per cent. between 1875
and 1896 and then, by 1908, shrank again by a quarter?
Yet that is what happened to the value of the gold
sovereign, measured in the amount of goods or services
it would buy.

When we have realized the underlying cause of general
agricultural prosperity or adversity, we can consider the
different surveys of the present position and the various
proposals for the future with a due sense of proportion.

Of the three official Reports we have named, that of
the Tribunal of Investigation covers the widest ground.
The Tribunal was appointed to consider the methods
adopted in other countries to increase the prosperity of
agriculture and secure the fullest possible use of the
land, and to advise as to the methods by which those
objects can be achieved in this country. This it has
done in a valuable survey, final majority and minority
reports being issued in May 1924.

Since 1870 the area of land under the plough at home
has diminished by three or four million acres, but abroad,
by the adoption of definite supporting policies, the arable
acreage on the whole has been maintained. Continental
countries rely chiefly on family farms, which we should
regard as smallholdings, while, in comparison, British
farms are larger, and more dependent on hired labour,
which receives better wages than in the rest of Europe.
Co-operation is more developed, not only in Denmark,
which farms for export, but also in Germany and
Belgium, which, like England, import food.

The differences are due to economic causes, and not

to the want of agricultural skill sometimes alleged.
Sir William Ashley and Professor W. G. S. Adams in
their majority Report say that this country "still presents some of the best farming in the world and is
unsurpassed as a source of breeding stock," while Professor Macgregor, who presents a minority Report,
shows that acre for acre the corrected yield of British
crops is as high as that of any other country but Belgium.
This is a very different picture from the one drawn by
some irresponsible critics.

In their recommendations, both majority and minority
Reports advocate development of agricultural education,
research and co-operation, an increase in the number
of smallholdings, and facilities for credit, as well as some
things which, like District Agricultural Wages Boards,
have since been established. Professor Macgregor looks
for these objects to be accomplished with less govern-
mental control than is contemplated by Sir William
Ashley and Professor Adams, who recommend also
direct State action in other constructive measures. They
hold that "under a Free Trade system Great Britain
can only maintain its tilled area by going over to arable
stock farming." They believe that the "disadvantages
attaching to any further decline in the arable area, will
be so grave that it will be worth while for the country
to pay a substantial price for its maintenance." But
they recognize the fact that the nation will only consent
to protect agriculture by import duties, subsidies or
guaranteed prices, and shoulder the burdens involved,
if it is converted to the view that broad national
advantages would thus be secured. Professor Macgregor,
on the other hand, thinks that military needs alone,

pressed on the Government by the responsible authorities, could justify us in paying the cost. Stabilization of the general price level, which would involve the whole subject of monetary policy, he regards as beyond the scope of the Report.

Lord Linlithgow's Committee was given the more restricted task of considering the distribution and prices of agricultural produce. Their final Report was issued in November 1923. After dealing with the general rise in price of agricultural produce and requirements during and after the war, a rise which they recognize as "largely due to increase in the currency," and the subsequent fall, they pass on to consider the spread between producers' and consumers' prices, a phenomenon so annoying both to the farmer and to the housewife. The Committee made an extensive study of methods of marketing and distribution, and set forth much detail of costs and charges previously unknown.

While recognizing the essential nature of the services rendered both by wholesale dealers and by retail distributors, the Committee reported that in many cases "the spread between producers' and consumers' prices is unjustifiably wide." During the time of falling prices which began in 1920, even good farmers made losses and farm labourers suffered a fall in wages. But distributors maintained high profits, and paid high wages. It is unjust that producers should bear the whole burden of a depression.

With regard to the future, the Committee recommend the stimulation of co-operative marketing by facilities for credit, improvement in the collection and the broadcasting of market intelligence, and of transport arrangements. They hope for a "development of a marketing sense" in farmers, and a standardization of agricultural produce. By such means the farmer may be put in a stronger economic position relatively to the dealer.

With these facts and recommendations before them, our three political parties have produced agricultural policies of their own. The Labour Party wishes to nationalize the land and place the import and distribution of food under some form of public direction[1]. The more extreme section apparently hopes to establish collective control of agriculture also, either by the State or the manual labourers. A Committee of Liberals carried out an extensive enquiry of their own, and have put forth the results in a bulky volume of 584 pages[2]. They advocate the nationalization of land but not of agriculture, a scheme which, with modifications, was adopted by a Conference held by the National Liberal Federation in February 1926[3]. Some Conservatives support protection and some subsidies, while most wish to see a great increase in the number of occupying owners. Finally, the Government, weighted with the responsibility of office, has issued the White Paper of a blameless line of policy, which, while containing much that is common to all sides, pleases none of the enthusiasts of the contending factions[4].

The White Paper renounces "drastic and spectacular

[1] *A Labour Policy in Agriculture*; published by the Trade Unions Congress and the Labour Party, July 1926; approved by the Labour Party Conference, see *The Times*, October 14th, 1926.

[2] *The Land and the Nation*; Rural Report of the Liberal Land Committee 1923–5; Hodder and Stoughton.

[3] *Report of the Land Conference*; Liberal Publication Department, London, 1926.

[4] *Agricultural Policy*, 1926, Cmd. 2581.

action on the part of the State," and in its positive proposals is perhaps a greatest common measure of the other
pronouncements. It promises support of the existing
Wages Committees as the best means of securing the
highest wages the industry can afford, and a continua-
tion of the work of the Forestry Commission, of agricultural research and education, and protection from
animal and plant diseases. It announces new facilities
for agricultural credit, for rural housing, for cottage
holdings and smallholdings, for land drainage, and for
co-operative marketing. Where legislation is needed,
the Government is already setting out to redeem these
pledges.

Probably all political parties would approve the
objects thus set forth. Though we may expect some
criticism of the methods proposed for carrying them
out, in substance the necessary action should secure a
considerable measure of agreement.

Turning to proposals for more fundamental change,
we reach controversial ground. A subsidy would benefit
any trade, and certainly protection would help one
that supplies only the home market and suffers
from foreign competition. A stronger case can be
made out for agriculture than for any other industry.
During the war there was a very general feeling that
agriculture should never again be allowed to fall into
such a depressed state as it did from 1880 to 1895, and
that national interests called for some kind of guarantee
to the farmer[1]. But the guarantee given in the Corn
Production Act of 1920 was repudiated as soon as the

[1] See the *Report* of the Agricultural Policy Committee of which
Lord Selborne was chairman, 1917, Cd. 8506.

cost to the Exchequer was realized. Personally, though I shared the general view during the war and supported a guarantee, I recognize that times have changed. I now think that, in present conditions, the balance of national advantage inclines against protection or subsidy, for reasons that will appear below, and, whatever view we may take as to their desirability, we must admit that, in present political circumstances, they are impracticable. This conclusion has now been accepted by a Conservative Government and endorsed with even greater emphasis by the two other parties.

An extension of occupying ownership is warmly advocated by some who believe that "the magic of property turns sand into gold," and the example of Denmark is often adduced. Many English tenants bought their holdings after the war, to prevent their homes from being sold over their heads. But there is no general desire among English farmers to sink capital in land. The economic state of Denmark—a small and comparatively poor country in which agriculture is the chief industry—is so different from our own that comparisons are misleading, and there is definite evidence from America, where both systems are common, that agricultural efficiency and output do not depend so much on tenancy or ownership as on other factors.

Some of those who have realized that the determining cause in agricultural prosperity is the price of agricultural produce have gone far beyond the modest desire of the Government to help co-operative marketing with credit facilities. They have not faced the fundamental problem of the stabilization of the general price level, but they have worked out a complete scheme for national management of the import and distribution of food, and the artificial control of agricultural prices, both to producer and consumer. They might thus succeed in smoothing out short-term fluctuations in price, but their scheme would leave untouched the long, slow drift of prices on which general agricultural prosperity or depression depends.

This policy of partial stabilization was suggested by Mr Bruce, the Prime Minister of Australia, at the Imperial Economic Conference in 1923; it has received sympathetic reference both from Mr Baldwin and from Mr Snowden in the House of Commons; it was described in the *Report* of the Committee on the Stabilization of Agricultural Prices, and has been adopted in the Labour Pamphlet. It is dealt with in a short section of the Liberal Land Book, only to be rejected. Its dangers are doubtless real, and are those common to all extension of collective trading. Nevertheless, many enquiries show that there is too great a difference between the prices paid to the farmer and those charged to the final purchaser. Moreover, fluctuations in price are the greatest obstacle to successful agriculture, and its stabilization at a remunerative level more to be desired than any other reform. Even the elimination of short-term oscillations would be a real advance. Much experience in public management was obtained during the war, and the success of many existing large trusts and combines indicates the possibility of extending their methods. Proposals for the collective control of dealing and distribution are by no means confined to the Labour Party. We shall consider below its advantages, drawbacks and difficulties.

Both the Liberal Book and the Labour Pamphlet advocate the nationalization of the land and the expropriation of the private landowner, in this carrying on the ideas of many socialist writers from the latter part of the eighteenth century to our own day. On more practical grounds than the early visionary idealists, our two parties of ardent reformers agree in desiring that the nation should take possession of the land, either at once, as advocated in the Liberal Book and the Labour Pamphlet, or gradually, as preferred by the Conference of the Liberal Federation.

Doubtless, some of the weight of this attack on private property in land is due to political prepossessions, to the long-standing Liberal dislike of the country landowner, and the more recent Labour dogma of public ownership of the means of production. But the expropriation of the rural landowner is also advocated by Mr C. S. Orwin and Colonel W. R. Peel, of the Oxford Institute of Agricultural Economics[1]. Their book is entirely friendly to the landowner and favourable to the system he has represented. They have come, evidently reluctantly, to the view that the present system is breaking down owing to the financial pressure on owners, which has led to the sale of so many estates during the last few years. Clearly such an opinion needs the most careful and respectful consideration. It is put forward by acknowledged authorities, free from any taint of political bias.

In the course of our survey we shall have much to say about this proposal to nationalize land. Mr Orwin and

[1] *The Tenure of Agricultural Land*, Cambridge University Press, 1925.

Colonel Peel have explained their reasons and their conclusions in a slim and pleasant volume of 76 pages. The
Labour Party has, probably with wisdom, refrained from
weakening its decision by giving evidence. Perhaps
there was no room for it in a pamphlet. The Liberals,
on the other hand, with greater courage, have written
a book. Here, at any rate, are given details which can
be subjected to critical examination. We can see what
is the value of the evidence on which this school of land
reformers relies, as well as what it wishes to do, and
what the democratic nationalization of the land means
in practice. Indeed, it is clear from Labour speeches
made in all parts of the country, that Labour orators
are taking the information and conclusions of the Liberal
Book as the basis of their own land policy. The book
will have much influence in coming years.

In *The Land and the Nation*, sometimes known from
its cover as the Green Book, the Liberal Land Committee describe the present state of British rural life and
agriculture as it appears to their investigators, and work
out a scheme whereby they think its troubles may be
alleviated or ended.

The Committee hold that, while some British farmers
are skilful agriculturalists, the majority are not making
the best use of their land, the proportionate output in
Great Britain being less than in some other European
countries. In spite of costly efforts, the number of smallholdings is getting less, and England remains chiefly a
country of landless labourers, dependent solely on wages
for a living. Owner occupiers are more in number than
before the war, but three-quarters of the cultivated land
is still worked by tenant farmers, whereas in other

countries ownership in some form predominates. Hence
the conclusion is reached that our system of land tenure
is at fault. The landowner, it is said, has ceased to lead
in agricultural development, and can no longer afford
to find adequate capital for the equipment of the land.
He must be expropriated. The same train of argument
is accepted as established in the Labour Pamphlet and
at the Labour Party Conference which adopted it as the
official Labour policy. Hence, in examining its validity,
we are dealing with the proposals of both these groups
of land reformers.

The Green Book points out that in the mediaeval
manor the lord held by military service, and every sub-
tenant had both rights and duties on the land. By the
decay of the manor and the growth of enclosure has
arisen the modern system to which the book gives the
bad name of "landlordism." To this all evils are traced.
To cure these evils, the State must now resume possession of the land and require adequate cultivation,
the modern equivalent of feudal dues, from those to
whom it leases it. They will be given security of tenure
at fixed rents in return. When the change is effected,
the administration will be entrusted to a new, democratic, County Agricultural Authority, partly elected,
and partly nominated by the Ministry of Agriculture
after consultation with the local unions of farmers and
labourers.

It is interesting to find the Liberal Party, even though
it be somewhat late in life, converted to a belief in the
benefits of the feudal system. Again, we seem to remember Mr Lloyd George railing at landowners because they
were rich men who toiled not nor spun. It is therefore

a source of some pleasure to find them now blamed for being poor by the party which invented death duties and made them so.

But *The Land and the Nation* is not to be laughed out of court. It contains the results of much hard work; its criticisms are sometimes justified, and are generally better than the ignorant nonsense about rural matters that often proceeds from urban, or suburban, sources; it contains an elaborate and, on its own unsound lines, an ingenious scheme for a new system of land tenure, and, above all, while proposing to nationalize the land, it abjures the hopeless idea of nationalizing agriculture as advocated by some Labour reformers who wish agricultural workers of whatever grade to be salaried servants of the State with no pecuniary interest in the success or failure of their operations.

Now expropriation by the State would solve a difficult problem for those owners who wish or are obliged to sell agricultural land of little residential value. At present, few or no farm tenants will buy willingly, and often the only purchaser in the market is the land speculator who buys an estate in one block to sell it again in many, putting pressure on the tenants to buy at high prices by methods the ordinary landowner cannot or will not adopt. Personally, having no objection in principle to public ownership, I would far rather be bought out by the State than be obliged to sell to a land speculator.

But, after all, there are still many country landowners who do not wish to sell their family property, and see their way to carry on; there are still, as in past ages, men of means, willing to buy land, that they and their descendants may enjoy its amenities and shoulder its responsibilities. I believe it will be well for the nation to allow them to do so. I venture to think that even the Oxford economists regard too much as permanent some of the difficulties that are temporary. I hope to show that most of the Liberal and Labour arguments are based on a partial reading of present indications and an incomplete appreciation of the fundamental economic and social causes which have produced them. I think that the terms of expropriation both as set forth in the Labour Pamphlet and the Green Book are unjust to landowning families. This was admitted as regards the latter at the Conference of the Liberal Federation, when fundamental amendments were accepted. I believe, further, that the scheme of administration proposed would involve the State in heavy financial loss, and that it would fail to produce the beneficial effects on agriculture and rural life that its authors desire.

Most of the remedies summarized above for the cure of agricultural ills have been put forward without a systematic study and a correct diagnosis of the disease to be treated. It is true that farmers have had four or five difficult years of falling prices and are sowing some land down to grass. It is true that, owing to the losses they have suffered, the wages they can pay to agricultural labourers are less than is desirable. It is true that the rents of agricultural land have not risen in anything like proportion to the cost of repairs and maintenance, and that landowners for the time are impoverished.

But agriculture has been through worse times of depression before and recovered. Moreover, the investigations of economists have now clearly revealed the causes of depression and recovery, though few of our would-be land reformers seem to be aware of this, the most important element in their problem.

As stated above, while the price of individual crops may oscillate with plenty or scarcity, the broad changes in the price level, which bring general agricultural prosperity or adversity, occur simultaneously in agricultural produce and in other commodities. For the cause of these changes therefore we must clearly look, not to factors which concern one industry only, but to those which affect all, to the general economic and monetary conditions on which price depends.

Those who read this book will find reasons given for rejecting most or all of the more "drastic and spectacular" measures which have been proposed. Nevertheless, if I criticize the conclusions of Conservatives, Liberals and Labour men alike, I gladly recognize the desire for the improvement of rural life which animates them all. Indeed, could they cast aside their respective obsessions about the iniquity of foreign competition, of the country landowner and of the capitalist system, I think they might meet on much common ground. They might even join to face the fundamental problem of stabilizing the general level of prices, and meanwhile carry out modest reforms which singly may seem insignificant to enthusiasts, but together would do much to enrich both the economic and social life of the rural community, sadly disintegrated as it is by the undirected changes of the last century and a half.

In return, I venture to hope that the following pages may be taken as a contribution to the solution of a difficult problem, and an analysis of the causes underlying our troubles, necessary before heroic remedies are adopted.

# PART I

## *BRITISH AGRICULTURE*

### Chapter I

### ARABLE AND GRASS LAND

THE first and chief point made in the writings and speeches of most would-be land reformers, is that the output of English land is less than it should be, and less than the yield in other countries, owing to the want of knowledge or skill of our farmers, and to the system by which they rent land from landowners who are for the most part impoverished and inefficient. On this statement the argument for radical change rests, and, if the statement can be shown to be based on a misapprehension of facts, the whole case falls.

That the gross output of food from English land is less than from an equal area in some other countries (*e.g.* Denmark or Belgium) is undoubted. Sir Thomas Middleton's pre-war study of German agriculture (often quoted and nearly as often misunderstood) proves the same thing for that country. But we must not fall into the common error of arguing direct from these figures of gross yield to bad farming. The chief reason for the difference is the smaller proportion of English land under the plough. Arable land can always produce more food per acre than permanent grass, and employ more labour, and for this reason most independent enquirers and all political parties desire to increase the proportion of

2-2

plough land. But, while the Conservatives see that this can only be done by increasing the price or diminishing the cost of production of arable crops, the Liberals hope to effect it by tinkering with our present system of land tenure. The Labour Party, by the more grandiose scheme of nationalizing both land and the import and distribution of food, might, at an unknown cost to the Exchequer and the nation, at least put up prices to the farmer.

This question of arable and grass land lies at the root of agricultural economics and politics, and it is necessary to deal with it at some length at the outset of our enquiry. Why does the British farmer prefer grass farming and why has the proportion of grass increased in this country since 1872, except for the short-lived "ploughing campaign" in the latter part of the war?

There is a long history of controversy, religious, social and political, behind this subject. But in present conditions, especially when agriculture is hard pressed, it must be chiefly an economic problem. Methods of agriculture are perhaps slow to adjust themselves, but the amount of arable land will tend towards that quantity which gives to the farmer the best return in total satisfaction, reckoning in and balancing against each other, money receipts and outgoings, pleasure in providing employment and the anxiety of undue financial risk.

The chief factors are climate and soil, on the one hand, and prices and costs on the other. In the moist climate of the West of England, heavy clay cannot be cultivated under the plough—the trouble and expense are always prohibitive, and during long stretches of time the land is too wet to be workable. Large areas of light soil in the drier districts of the Eastern Counties, and still more on the Continent of Europe, will not carry good grass permanently. Much of this kind of country is almost unfenced, and water for stock is in places unobtainable. But between these two extremes lies a certain amount of English land, perhaps four or five million acres, which can be put under either grass or arable crops as the economic circumstances vary one way or the other. When prices rise or costs fall, it pays to plough up grass and get the higher yield of arable land. If prices fall or costs rise, this tendency is necessarily reversed; the greater risk and expense of arable farming make it unprofitable, and, if farmers are to remain solvent, some land must be sown down to grass. The dependence of the area under different arable crops on the price of the produce is well known—it is illustrated by the tables given in the *Report of the Agricultural Tribunal of Investigation*[1].

Thus, not only does the drier climate of the Continent make for more arable cultivation, but economic factors also are involved. First, the cost of labour is lower abroad. The Agricultural Tribunal, after careful investigation, found that, even before the war, the real wages of farm labourers in this country were higher than in Denmark or in Holland, and higher by 25 to 40 per cent. than in Germany, France or Belgium. Moreover, Danish labour at all events is reckoned by competent observers as considerably more efficient than ours. The only consideration to be put on the other side is the

[1] *Report*, pp. 301–6.

fact that English rents are lower than in most other countries. But, on balance, the cost of large-scale arable farming is higher here than it is abroad. And, on the small scale, the number of men ready to face for themselves and their families the life of unremitting toil of the arable smallholder is less. Furthermore, in England, a comparatively rich country, there are more profitable ways of employing brains, capital and labour than in working agricultural land. It pays better to invest in picture palaces, tobacco or cocoa than in land improvement or farm stock, and better to become a bricklayer's assistant or a municipal dustman, or even to remain a hired farm labourer, than to work as hard on a smallholding as does a Dane or a Belgian. Thus the point at which arable cultivation ceases to pay is sooner reached in England. It is only the fortunate fact that we can grow good grass which has saved much of our land from becoming derelict, and surely we need not regret this our good fortune.

In the comparison between arable and pasture, the advantages are not all on one side, as is so often assumed. Of course it is true that the gross output of food, measured in starch-equivalents or in calories (the units of heating value), is two or three times greater from arable land. But the total financial receipts are about the same per acre, and, per man employed, are greater from grass farms, so that higher wages can be paid. Many enquiries confirm these statements. It will suffice to quote one of them. Mr H. J. Vaughan, comparing arable farms in the Cotswolds with grass farms near Rugby, on the average of large numbers, found that the annual production per acre was £6. 10s. from the arable land, and

£6. 7s. from grass, while the receipts per man were £261 on the arable as against £477 on grass[1].

The truth is that the British farmer, like the British manufacturer, finds it better to specialize in high-grade products. Milk and meat contain a large proportion of the more valuable and digestible proteins, and, where good pasture can be grown, the simplest and cheapest way of producing milk and meat is on permanent grass.

Furthermore, it should always be remembered that British pedigree flocks and herds hold a position of unquestioned supremacy. Other nations of the world come to us to establish and replenish their stock, and those of our breeders who have developed this industry have created a national asset of great value. Yet we may look long in the writings of our critics before we find a reference to this characteristic and successful activity of the British farmer, especially of the farming landowner.

In any other industry, such a balance of economic advantage as is given by grass land, with its better profits and higher wages, would be accepted without question. Indeed, in the new countries which supply the world with corn, the whole system is based on accepting a low yield per acre in order to obtain a high output per man employed. But in England the trouble is that agriculture cannot expand to absorb the labour displaced by improved methods. The area of English land is limited, and the world's demand for common foodstuffs is much less elastic than for manufactured goods. A man, or at all events a woman, can buy any number of new clothes and other amenities of life with increasing satisfaction,

[1] *Journal of the Royal Agricultural Society*, vol. LXXXV, 1924.

but there are natural limits to the amount of food he or she can consume. Hence not only here but in all countries there is a "drift to the towns," that is, the proportion of men employed in farming is continually falling with the increase of population and of wealth. This is inevitable; and, though we may regret some of its features, it is on the whole a healthy sign of a general rise in the national standard of life. If new industries could be established in rural neighbourhoods, or existing industries moved from the towns to the country to give the necessary employment, the chief cause for regret would be removed.

The only way the tendency of the people to drift away from agriculture can be checked is by the action of some definite agricultural policy which modifies the economic forces. Such policies are far more necessary in countries which cannot grow permanent grass—indeed their only alternative would be to let the land become derelict. Different possibilities are well shown by the methods adopted by Denmark on the one hand and by Germany on the other to meet the fall in prices produced by the shortage of gold from 1873 onwards.

Denmark maintained free trade, and, accepting the increasing supplies of American corn at the falling level of prices, used them as feeding stuffs to supplement home-grown fodder and thus feed the cattle needed for an expanding dairy industry. This system works chiefly for export to Great Britain, and is thus specially suitable for careful grading of produce and other co-operative activities. A uniformly high standard is maintained for export by using at home produce that is not up to that standard. Such methods could only succeed in a small country comparatively poor in alternative industries, whose whole surplus output is absorbed easily by a neighbouring and growing market freely open. When the new system was established, British dairy farmers were finding it more profitable to sell their milk, and thus left the butter and cheese trade more and more to be supplied from abroad. In this way Denmark was able to preserve her plough land by arable dairying. Some British arable farmers with suitable land and markets might well have followed her example. But our whole corn growing area was too wide thus to find salvation.

Germany also was too large a country to specialize in this way, and mitigated the fall in prices by import duties. The fall in prices was due to monetary causes which we shall deal with fully in Chapter v. Protection did not prevent it. But corn growing, though less remunerative, indeed sometimes carried on at a loss, was maintained better than in England, and the land kept in cultivation. This was done, of course, at the immediate expense of other industries. As Sir William Ashley and Professor W. G. Adams say in the majority Report of the Agricultural Tribunal of Investigation:

It is necessary also to realize that German statesmen decided to support agriculture by tariffs, even though thereby they should impose some check on industrial development. Among their serious thinkers there have been none to maintain that over a relatively short period, say of three or four decades, it is feasible to have the largest possible development of agriculture and also at the same time the largest possible development of industry. Some of them have indeed argued that, in some distant future, a secure agriculture will prove a securer basis for home industry than foreign trade. But, for the future within sight, they have granted that to ask the manufacturing population to pay somewhat more for their food than they could get it for from America or Russia, was to impose some restraint on the growth for export of German manufactures. They could meet more or less effectively the assertion that protective duties on food meant positive harm to the industrial population by pointing to the actual growth in German exports, and to the statistical evidence of improvement in the workman's remuneration. But they did not deny that exports might be larger still and urban remuneration even better if there were no duties on food. What their political and intellectual leaders, like Von Bülow and Professor Adolf Wagner, asserted was that it was worth while somewhat to slacken the progress of German manufacturing industry, if thereby other ends were achieved which they deemed more beneficial to the nation....

If it be granted, as we think it must, that German policy, and before all else its tariff policy, was for the benefit of German agriculture, two comments must be added. The assertions that Germany kept on the soil a larger population than England and obtained from it a larger production of food should not be taken to mean that German agriculture is more "efficient" than English agriculture; if by "efficient" we mean productive in proportion to cost. In that sense, there is certainly no superiority in German over English agriculture, except perhaps in certain special fields. Germany keeps a larger number of people working on the soil, and gets a greater gross output. This greater gross output is naturally all, directly or indirectly, made use of in maintaining them there. If Germany were to allow most of them to depart, and would be content with a smaller output, it might obtain a greater *net* output. But that has not been the object she has set before herself.

Now there is more to be said for artificially supporting agriculture than any other industry. Natural uncertainties are greater and stability of price more necessary, depression is more disastrous socially, since it leads to depopulation of the countryside. It is important to find a career at home for those of our people whose special aptitudes are suited to an open-air life. A contented and prosperous peasantry is a sound and wholesome element in the population. There is additional security in time of war.

Germany, seeing all this, paid the necessary price by checking the rate of industrial progress in a time of expansion. I think it would have been wise for us to have done likewise; to have let industrial development go on more slowly and kept it more under control; to have taken the best from the mediaeval conception of the social organism, and adapted it to modern needs. But it is far more difficult to face the cost when expansion has ceased, when, perhaps, contraction is upon us. However support be given, whether by subsidy or protection, other interests must suffer, as German protectionists acknowledged.

Since some industries—building, railways and such like—are sheltered by natural protection, it would be fair that they should help those forms of agriculture which are exposed to world competition. But farming is not alone in this, and demands for help would at once arise from iron, coal, engineering and other industries.

If help were given to all by way of subsidy, the drain on the Exchequer would be enormous, sheltered industries would become less prosperous and total employment slacken. If, on the other hand, all the unsheltered trades were protected by tariffs, those that work mainly for export and are most depressed would gain no benefit, the cost of living would rise, trade unions in sheltered industries would claim higher wages, and the relative position become much as it is now. Probably the only considerable effect would be to raise the general internal price level, which, in a country that must export to buy food and raw materials, might be disastrous.

Protective campaigns in England always jettison agriculture when they come into touch with the realities of politics, and thus in practice would increase the cost of what farmers must buy, while leaving their produce still exposed to foreign competition.

The difficulties of changing from a policy predominantly free trade to one predominantly protectionist, or *vice versa*, are very great, and, except in times of industrial expansion, so great as to be dangerous. It is wise, perchance, to suffer our present evils, whichever they be. There seems some chance, in the course of the next few years, of general stability in agricultural prices, for reasons set forth below. It is probably wise to wait in hope, and meanwhile, either with or without State assistance, to do what is possible to secure more orderly marketing of special kinds of agricultural produce.

Thus, in present conditions, the balance of national advantage seems to turn against protection or subsidy for arable farming. But the discrepancies which now exist between wages and other costs in the sheltered and the unsheltered industries, if not corrected in a more straight-forward way, may modify this conclusion. They are the effect of naturally protected monopolies in an otherwise free-trade system. To this point we shall return in a later chapter. We shall also consider at length the third method of bringing State help, namely, the national control of imports and markets.

Whatever be our opinions on the abstract question, there is no doubt about practical politics. Governments will not face the cost of heavy subsidies, and every time a tentative proposal of general protection has been put forward, the fear of dearer food has driven the electorate to reject it by overwhelming majorities. The present Conservative Government has now recognized that protection or subsidy, adequate to stay the shrinkage in plough land, is not possible in present conditions, and that little additional military security would be obtained by any increase in arable cultivation which could be brought about by such means[1]. The other two parties, even more dependent on urban votes, are still less likely to reverse this conclusion.

Therefore it seems that farmers will be wise to face the facts of the political situation, and give up hope that arable cultivation will be stimulated either by protection or by some large subvention from national funds. But, when they have done so, they may fairly claim that their political critics must in their turn face the facts of the economic situation, and not expect farmers to produce crops which in present conditions it cannot pay to grow. For instance, it must not be forgotten that every increase in rates of wages registered by the Wages Board, desirable though it be, means an increase in costs of production, unless, indeed, it is accompanied by a proportionate rise in price of produce or in efficiency of labour, or a decrease in rents already too low to secure an adequate flow of capital for the maintenance and improvement of the equipment of the land. By every uncompensated rise in wages or other costs, arable land on the margin is made unremunerative and must be sown down to grass, and bad grass land on the margin must become derelict.

[1] White Paper on Agricultural Policy, 1926, pp. 2 and 3.

And now let us sum up the results of our enquiry into the balance between arable and grass land.

There has been much confusion of thought in discussions on this point. People often profess to desire an increase in the prosperity of agriculture and the rates of wages paid, and also an increase in the number employed or settled on the land, as though both objects were identical. But the truth is otherwise. Except in so far as the economic conditions make it profitable to plough up grass land, high net profits and the accompanying high wages are usually incompatible with the employment of large numbers. As in other industries, total labour costs must be kept down if high wages are to be secured. To carry on a business with the primary purpose of creating employment is dangerous: it tends to defeat its own object. National wealth can only grow, and aggregate employment increase, if profits are made and some of them reinvested. To treat all industry as some wish to treat agriculture, would be a short and sure road to national ruin.

Let us keep the two aspects of the subject clear in our minds. Arable cultivation gives the greatest output of food and most employment. Profitable agriculture and high wages can best be obtained with grass land where good permanent grass can be grown. Where it cannot, the land must be put under the plough if it be economically possible. Arable farming will pay best and carry highest wages when conducted on the large scale with all modern improvements and machinery to increase the output per man. But, with labour-saving appliances again the whole amount of agricultural employment is diminished, though the additional national income will increase the aggregate employment in all industries taken together.

It is desirable to increase the output of home-grown food and also to provide a healthy country life for as many of our people as possible, but it must be realized that, if we do so in the face of opposing economic tendencies, we must pay the price in lessened national wealth and lessened aggregate employment. Probably the best compromise is to restrict direct national financial help to the organization of agricultural research and education, and to the development by appropriate forms of subsidy of new methods of marketing and new processes, such as the growth and treatment of sugar-beet, which cannot establish themselves unaided in the face of existing foreign competition.

But something may be done, by the extension of electric supply and in other ways, to encourage new industries to start in country districts and old industries to move out from the towns. This movement has an economic basis, and would proceed faster if the obstacles of inertia and prejudice could be overcome.

In this country it is probably wise to leave general agricultural operations to adjust themselves to the economic conditions of the time. To some extent economic conditions are under control. If Governments are to safeguard industry from renewed depression and agriculture from a further shrinkage in arable cultivation, they must face the difficult problem of stabilizing the general price level by international agreement and action. That would be a more effective measure than subsidy or protection. This fundamental point is dealt with in future chapters.

### CHAPTER II

## CROP YIELDS

HAVING shown that the smaller proportion of arable land in England is the inevitable consequence of climatic and economic causes, we must turn to the second point made by our critics, namely that, while some British farmers are skilful and successful, so many others fall short of a good standard that the average in general is low.

Now much of the evidence on which those who make this accusation base their case is in reality founded on the differences in gross output, which, as we have already seen, depend on the amount of land under the plough. But certain other evidence remains.

For instance, the Liberal book, *The Land and the Nation*, calls in witness the Agricultural Tribunal of Economists to prove that the "average yield of crops per acre shows a much larger increase over the last forty or fifty years in Germany, Belgium, Holland and Denmark than in Great Britain." There has also been "a shrinkage in the total area of our cultivated land, of arable land, of corn land, of roots and green crops, and in the number of livestock per hundred acres," whereas in some of the other countries named increases have occurred. Next, the Green Book adduces many instances of bad farming, especially where large areas are held by one man as sheep-walks or rough grazing, and gives extracts from the writings of Sir Daniel Hall, illustrating similar instances and summing up to the effect that "while we possess farmers full of enterprise...their example is not generally followed," and that, "farming is yearly growing less instead of more intensive." The conclusion drawn from all these facts and opinions is that, broadly speaking, British farmers are not doing their work well, and British land is under-farmed compared with that of other European countries.

Now let us take this evidence in order. Firstly, fifty years ago British agriculture was already extracting high yields from the land. Other countries have since followed our example and learnt to do so likewise. It has been easier for them to show an increase. Secondly, as regards the present relative position, one of the eminent economists on the Tribunal of Investigation, Professor Macgregor of Oxford, has himself dealt with the significance of the figures giving the present comparative yield of crops. In the *Economic Journal* for September 1925, he writes:

Suppose that the idea of productivity is considered in relation to even the crude yields of crops. Taking as 100 the average yield of Europe, exclusive of Russia, our indices before the War were, for wheat, 166; barley, 123; oats, 113; potatoes, 138. These indices would be much higher against the world averages. But a critical use of the crude yields is necessary, especially in the case of wheat, since other important countries grow rye for food, and obtain wheat on a smaller proportion of their cultivated area. If the indices of yield are weighted by the percentages of the cropped area under each of the crops for which there are returns—and these account for from 75 to 80 per cent. of the cropped area—then a composite index of productivity can be obtained, which makes Belgium 164, France 92, while Britain, Germany, Denmark and Holland are all on a par at about 130. The tabular statement and analysis are given in my Report to the Tribunal, and no criticism of the method or result has

appeared. The method is that used by the U.S.A. Department
of Agriculture in their Year-Book for 1919. The European
comparison for livestock is generally similar to that for crops.

Thus it appears that, except for Belgium, a land of
smallholdings, the productivity of British fields is as
good or better than that of any other country—even
of the much be-praised Denmark.

Thirdly, the estimates of the yield per acre of main
British crops, which on their face show only a very slight
increase or even a small decrease since 1872, need care
in handling. It would not be surprising had a real
decrease occurred. It is true that the worst land would
be given up first, and thus the average yield per acre
increased with decreasing area, but other factors work
in the opposite direction. With falling prices the law of
diminishing returns comes into play sooner, and intensive work ceases to pay at a lower level of cultivation.
Moreover, the land most difficult to work, which goes
out of cultivation soonest, the heavy clay, grows good
crops of wheat when it can be cultivated at all. Hence
the average yield of wheat in areas of heavy land, like
some parts of Essex, would naturally tend to fall in bad
times. On the other hand, Sir Rowland Biffen's new
and improved varieties of wheat should have caused
the average to rise during the last few years. It would
be difficult to predict which of these opposing tendencies
would be most powerful.

But Mr J. A. Venn has examined critically the methods
of estimation in a paper read to the British Association
in 1926, and published in the *Economic Journal* for
September of that year. He finds evidence to show that
the methods used have led to an increasing under-

estimate during the last fifty years, and that it is probable there has been a real and appreciable rise in the yield per acre of wheat, barley and oats. If so, in view of the difficulties encountered, it speaks well for the British farmer.

The decrease in the area of cultivated land, of corn, roots and other expensive crops, is due to the same causes which make the proportion of arable land less in England than in some other countries. In our climatic and economic conditions, corn crops pay their way on fewer kinds of soil than in countries where the weather is drier and costs of cultivation lower. As prices rise or costs fall, more land is ploughed up, more expensive crops are grown and more labour and fertilizers can profitably be applied, so that yields increase. When, on the other hand, as in recent years, prices are low and costs are high, nothing can prevent these movements being reversed. As long as agriculture is left free to face economic conditions, it is useless to complain in bad times of cheaper crops being grown, and the application of labour and fertilizers being restricted.

The fallacy that underlies the idea that falling prices may be met by higher farming is well known to practical farmers. It was exposed long ago by Sir John Lawes. And, indeed, if formal proof be needed, we cannot do better than quote the classical experiments on wheat at Broadbalk Field, Rothamsted[1]. Table I gives the average yields for the years 1852–64 with increasing amounts of manure.

---

[1] *The Book of the Rothamsted Experiments*, by Hall and Russell. See also *The Law of Diminishing Returns*, by Spillman and Lang (Harrap & Co.).

CROP YIELDS

The figures illustrate well the law of diminishing returns. Similar relations connect the yield and the amount of labour put into the cultivation of the land. On this point few experiments have been made hitherto, but the subject is now being investigated by the Institute for Agricultural Economics at Oxford.

TABLE I

| Plot | Manures per acre | Dressed grain | | Straw | |
|------|------------------|---------------|---|-------|---|
| | | Produce per acre in bushels | Increase for each 43 lbs. of nitrogen | Produce per acre in cwts. | Increase for each 43 lbs. of nitrogen |
| 5 | Minerals alone | 18·3 | — | 16·6 | — |
| 6 | Minerals and 43 lbs. of combined nitrogen | 28·6 | 10·3 | 27·1 | 10·5 |
| 7 | Minerals and 86 lbs. of nitrogen | 37·1 | 8·5 | 38·1 | 11·0 |
| 8 | Minerals and 129 lbs. of nitrogen | 39·0 | 1·9 | 42·7 | 4·6 |
| 16 | Minerals and 172 lbs. of nitrogen | 39·5 | 0·5 | 46·6 | 3·9 |

The effect of each successive dose of fertilizers or labour may increase for a time at first, but ultimately each effect becomes less than the last. The point at which additional treatment ceases to pay must obviously be reached sooner as the cost of labour or fertilizers rises, or as the value of the product falls. This result is illustrated graphically in *The Book of the Rothamsted Experiments*. It is, indeed, evident from the figures.

The extra 1·9 bushels of grain and 4·6 hundredweights of straw might be just worth getting at the cost of the third dose of 43 lbs. of nitrogen, if wheat were at 100*s.* a quarter and sulphate of ammonia at £12 a ton. But they will certainly not be worth getting if wheat is much cheaper or sulphate of ammonia much dearer.

The change from remunerative to non-remunerative expenditure is very sudden in this particular case.

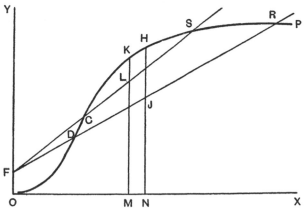

Diagram I. Law of Diminishing Returns, curve *OKSP*.

*OX* = amount of labour, fertilizers, etc. *OY* = monetary values.
*OF* = fixed charges.

Clearly, it is usually worth while at ordinary prices to give wheat land similar to that at Rothamsted somewhere about 86 lbs. of combined nitrogen per acre, and this optimum quantity will only be slightly affected one way or the other by varying prices or costs. But, with the total expenditure on cultivation including labour we should expect no sudden change. Allowing for possible increasing returns at first, we get a curve connecting $X$, the total treatment expended, and $Y$, the value of the resulting crop yield, more or less resembling the curve $OKSP$ in Diagram I.

The cost of treatment is proportional to the amount done, so that the cost curve $FJR$ for given rates of wages, etc., must be a straight line, starting from $F$, the point corresponding to the fixed charges on the land, comprising rent, rates, etc. The net profit is represented by the difference in height of the two curves $OKSP$ and $FJR$, giving receipts and expenditure respectively, and if this difference be measured, it will be found to be at a maximum somewhere near $HJ$, that is, at an amount of labour and other application of capital equal to $ON$, costing a sum represented by $NJ$.

Now let us imagine that prices fall or costs rise. We shall then get a new relation between the two curves, such as that shown by $OKSP$ and $FLS$. The maximum profit is now somewhere about $KL$, at an application of labour, etc., equal to $OM$—considerably less in amount than before, though possibly, owing to the rise in rates of wages, etc., costing as much or more[1].

Since the point $C$, where the line $FLS$ crosses the curve between $O$ and $K$, is to the right of $D$, where the line $FJR$ crosses it, we see that more work must be done on the land at the higher level of costs before the fixed charges are covered and any profit is made. At the other end, since $S$ is to the left of $R$, the profit will sooner disappear again if cultivation be overdone. With the higher costs or lower prices, the limits $CS$, within

---

[1] As long as the increments in yield diminish, the curve is convex upward, and it can be proved mathematically that $M$ is to the left of $N$.

which any profit is possible, are narrower than $DR$, the corresponding limits at the higher prices or lower costs. If a ruler, pivoted at $F$, be moved upward, this contraction of limits is well seen. If prices are low and costs high, it needs great skill to put neither too little nor too much work into the land to keep within the limits of a possible profit.

The right point at which to aim for each level of prices and costs needs careful estimation. The good farmer will judge it rightly, even if he does so instinctively. His land will always be cultivated as well as the times allow, but he is distinguished from the bad farmer by the greater effect he gets from a given expenditure on labour or fertilizers rather than by the mere amount of either that he puts into the soil.

One aspect of these results is illustrated by the fact that in new countries, where fixed charges are low and wages very high, it pays best not to attempt intensive cultivation. The average yield of wheat in Canada is 11·4 bushels per acre as compared with 32 bushels in England, though the output per man is greater in Canada. It is worthy of note that it is these new countries which really compete with English corn-growers, and not the countries of Europe with low rates of wages and more intensive cultivation, often held up to us as models.

If our would-be land reformers would "farm against their own cheque books" for a few years of falling prices, we should hear no more complaints against farmers because just now agriculture is becoming less and not more intensive. The remedy for bad prices is not to be found in forcing increased yields at a cost disproportionate to the value of the augmented returns. When a factory does not pay, it can work short time or be closed down till trade improves. A farm cannot close down; its operations have to be planned years ahead, and the land, if not kept in order, will become ruined by weeds, gorse, or thorn scrub. All a farmer can do in bad times is to cut down expenses while keeping his land in good heart, in the hope of better days. This is clearly the right, indeed the only, policy, but it may involve reducing somewhat the intensity of farming for a while, even to the extent of sowing land down to grass. However reluctant he may be, a farmer may have to economize on labour, but that is not his fault. It is the fault of falling prices due to such causes as post-war deflation, or to unavoidable economic changes—perhaps a decrease in the output of gold from the mines of the world. The trouble can only be prevented from occurring periodically by some such measure as the stabilization of prices. To this important point we shall return.

Extracts are frequently given by critics from the writings of certain agricultural authorities, especially from those of Sir Daniel Hall, to show that British farming is not all it might be. It seems unfair to act on a theory of verbal inspiration, and copy sentences divorced from their context, as do some of these critics. But if we are to quote texts from the Books of Sir Daniel, as many can be found on the other side. Here, again, let us turn to Professor Macgregor, who is neither a landowner nor a farmer, and allow him to choose our texts for us:

In two ways the statement that we are seriously under-farmed can be considered—the direct observation of experts,

and the use of the returns of yields. As to the former, both
Sir Daniel Hall, in his *Pilgrimage of British Farming*, and
Sir Thomas Middleton, in his *Food Production in War*, say
the same thing; that farming, like any industry, can be
improved, but that it is as well served as any other industry
or profession. Let anyone compare with the phrases of
political discussion the summary chapters of Sir D. Hall's
*Pilgrimage of British Farming*, or such local comments as
that in the West (? East) Riding flats, "agriculture is a good
driving business, which is getting out of the land something
approaching the highest yield that is profitable"; that, in
Northumbria, "only a determined and skilful race of farmers
could attain the prosperity of which we saw so many signs";
that, in the Lothians, "we had not imagined that the
management of arable land could reach such perfection";
that, "a study of the Shrewsbury district may be recommended to those who declare British farming to be a lost
art"; and others to the same effect. Had this been written
of the districts of a foreign country, it would have been
quoted in England in italics.

Such is the verdict of an independent and expert
enquirer on the practical craft of British farming. And
we have as little need to fear comparison in scientific
agricultural research. The great "improving landlords"
of a century ago are connected with modern investigators through the work of Sir John Lawes at Rothamsted, which represents both the estate of an "improving
landlord" and a modern experimental station. There
the chemical principles of manuring were established,
and there the latest physical, chemical and biological
studies of that complicated organic structure we call the
soil are carried on. Cambridge has made striking advances
in plant breeding, genetics and animal nutrition. Oxford
has helped to teach us how to treat grass land and is the
headquarters of agricultural economics and agricultural

engineering, and the various Agricultural Colleges about the country are making worthy contributions to knowledge. The Royal Agricultural Society is subsidizing research and assisting the publication of results, while the Ministry of Agriculture, advised by most competent and helpful experts, is a model to some other Government departments.

The weakest link in our agricultural chain has been the method of making known the results of research in a practical and helpful form. But every year the County Agricultural Organizers are getting into closer touch with farmers, and every year farmers are learning to rely more on them and on the advisory staffs of the various Schools of Agriculture for advice. The Royal Agricultural Society is now publishing an annual volume, in which recent research is described and its practical bearings made clear. We may fairly claim that our modern agricultural science is worthy of our great traditions of agricultural practice.

Finally, let us deal with the question of the size of the holdings. The fact that British farms are, on the average, larger than those of other countries, seems in itself to be an offence in the eyes of some land reformers. If British factories were shown to be larger than foreign ones, would they propose to cut them up?

Some forms of agriculture, especially the arable cultivation it is wished to encourage, can only be carried on successfully on a scale large enough to employ expensive machinery and pay high wages, both for management and specially skilled labour. The scientific and progressive farmer can only find adequate scope for his abilities, and contribute his full addition to national wealth and welfare, if he is in control of a large area of land. It is abundantly worth the nation's while to give him full opportunity.

Smallholdings are useful to provide a ladder for exceptional men, and all parties are pledged to a policy of increasing their number. But the successful smallholder and his family work harder and for longer hours than the "landless labourer," and often for a smaller return. Unless England becomes a poorer country and English wages are lower (which may Heaven forbid!) it is hopeless to expect that most of our land should be cultivated on the model of Belgian smallholdings. Cottage gardens, and cottage holdings for labourers working also for hire, are another story. They are compatible with large farms, and produce social as well as economic benefits.

Figures given below, it is true, go to show that the uncorrected gross output per acre from individual farms increases as the size of the holding diminishes. Hence people argue that, by subdividing the land, the total net production of food would be raised. They overlook the fact that the figures of gross output take no account of foodstuffs purchased. The smallholding is less self-contained than the large farm, which grows most of its own requirements. If uncorrected figures are used, the gross output per acre would be greatest from a market garden with ranges of glasshouses, or from the buildings and yard of a town dairyman, who keeps his cows in a disused stable, and buys all the food they want.

Moreover, the smaller the holding the more labour is needed per acre, so that the output per man increases with the size of the farm. This is well shown by figures quoted by Mr Orwin from Mr Pryse Howell[1]:

| Acreage of farm | Production per acre | | | Production per man | | |
|---|---|---|---|---|---|---|
| | £ | s. | d. | £ | s. | d. |
| 1– 50 | 11 | 19 | 9 | 168 | 19 | 0 |
| 50–100 | 9 | 19 | 2 | 156 | 2 | 0 |
| 100–150 | 7 | 19 | 1 | 189 | 0 | 0 |
| 150–250 | 7 | 5 | 8 | 222 | 12 | 0 |
| Over 250 | 8 | 4 | 4 | 316 | 19 | 0 |

These results show, with regard to the size of farms, the same kind of antithesis as appears between arable and grass land. Small farms give a greater total output of food per acre, and employ more labour counting in that of the occupier; but large farms give a greater output per man, and can therefore provide a higher standard of life.

When we pass from gross output to net profit, we are met with the difficulty of assessing the value of the labour of the holder and his family on small farms. Nevertheless, there seems some evidence to show that efficiency is greatest (1) on farms of between 75 to 150 acres, which probably represent the economic unit of a family farm, fully using one pair of horses, and (2) on farms of about 350 acres, probably indicating the area of land which the average farmer, working with hired labour, is best able to control. Exceptionally able men, of course, only find adequate use for their powers in businesses of much larger size.

Of course, if all farmers were good farmers, the yield of British land would increase. So would the output of

[1] *Journal of the Royal Agricultural Society*, vol. LXXXIII, 1922.

British factories if all manufacturers were as skilful as the best. So would the benefits of British politics if all politicians were as able and scrupulous as they are loquacious. It is an imperfect world, and the amount of ability therein is limited. We shall always find differences in performance, and, especially in agriculture, where success depends largely on attention to detail, on a knowledge of the peculiarities of each field, even of each animal, it is impossible to standardize practice, and differences in results must be considerable.

There is nothing new in most of the evidence we have adduced. It has been available for months, and has been brought again and again to the attention of those who began this attack on the management of British farming. Some of them pay no heed to facts and figures, but repeat their original complaints.

Misstatements may be made once owing to honest ignorance, but what are we to say of those who repeat them when refuted by definite, quantitative, scientific proofs, given with facts, calculations and references by impartial experts? Let us once more quote from Professor Macgregor, who says[1]:

Economists were asked to give a judgment on all the aspects of the problem when the Tribunal was set up. They strove to separate the issues and to show the nature of the choice that is before the country. But what use is it all if no Enquiry will be acceptable which does not supply the superlatives that will give *éclat* to a political drive? And especially if the policy is to be based on the sheer libel that British farming with its great history and its great contributions to agricultural method, is to be discussed in the company of such phrases as "the worst," or even "among the worst" in the world?

[1] *Economic Journal*, September 1925, p. 397.

### Chapter III

# LAND IMPROVEMENT AND RECLAMATION

THE idea that the production of food should be increased regardless of cost is one that most land reformers would scout, yet it underlies much of their argument. Take, for instance, the following extract from the Liberal Green Book[1]:

"Perhaps the most striking example of what can be done with English land is derived from the Midlands, where an Englishman took over some years ago 500 acres of land to reclaim, which at the time was yielding no income to the landlord, except for shooting, and producing no crops. It is now carrying a considerable number of stock, cattle, pigs and poultry, producing valuable crops and paying an annual wages bill of over £1000." The tenant started with a capital of "one thousand pounds raised through the bank and guaranteed by his two brothers." The landowner, though "he contributed nothing to the success of this very striking bit of reclamation, was, at any rate, not in any way obstructive."

Now, unfortunately for the Liberal Committee, this land was identified, and correspondence appeared in *The Times*[2] which showed that: (1) the landlord spent over £6000 on houses, buildings, etc., and (2) the Ministry of Agriculture took over the farm in 1918, worked it through the local Agricultural Committee for three years, and expended not less than £10,000 on reclamation. The recklessness of statement thus revealed does not

[1] *The Land and the Nation*, p. 115 *et seq.*
[2] October 21st to October 29th, 1925.

speak well for the general care and accuracy of the authors. But a more serious point remains. Over £16,000 was spent on this land besides that contributed by the tenant. Even then further expenditure was said to be needed on reclamation, buildings and fencing before the farm could be satisfactory. After the work was done, the rent was £286. 6s. 8d.

Suppose we reduce the £10,000 spent by the Ministry by one-half, to allow for the extravagance of public administration and the high level of costs at the time. We still have an expenditure of £11,000 to produce an increase in annual value of £200 or £300. Whoever owns the land, whether the most grasping of landlords, the most worthy of occupying owners, or the most democratic of County Authorities, would find many such proceedings a short road to ruin.

Of course this is an extreme case; better results might sometimes be shown by less ambitious and more economical schemes. But land improvement at present costs is rarely remunerative, even where drainage alone is needed. The cost of pipe draining whole fields anew is now quite prohibitive, running up to amounts which equal or exceed the capital value of the land. But sometimes old drainage systems are not working only because the outfalls are choked, and, if the position of these is known (usually no plan exists), a moderate expenditure sets the old drains at work again. Then, if heavy land is uniform clay, the modern method of mole draining is both cheap and effective. A vertical knife, carrying at its lower end a horizontal cylinder a few inches in diameter, is pulled through the ground by a windlass or winding engine, leaving a free channel bored in the clay, a channel which will remain open for many years. Much land is being improved in this way, but unfortunately some clay land has patches of gravel or big stones scattered about, which may make it impossible to draw a mole. Except where old pipe drains can be repaired or the mole plough used, no large schemes of drainage can pay for the expense of carrying them out.

The other chief cost of improvement is the erection of new buildings, or the reconstruction of old ones. This is much more important, because it applies to all farms, whereas drainage is only needed on heavy land.

The expense of building has increased out of all proportion to the rise in the general level of prices, that is to say, to the fall in the value of money. The need of the country for houses has been used by the building trade to exploit a jealously protected monopoly. Tribute has been exacted from the nation in the form of subsidies, the cost of work such as new farm buildings, not eligible for subsidy, has been made almost prohibitive, and the repair of old ones a very heavy burden. Materials have risen in price, but on the whole not more than the increase in labour and other costs warrant[1]. It is possible that some master builders are making unreasonably high profits. It is certain that the operatives are obtaining weekly wages which have increased over pre-war figures by 90 to 125 per cent.[2] In addition, hours are somewhat shorter, and output, at all events of bricklayers, much diminished—a minimum of 750 bricks laid in a day before the war having now fallen to an average which

[1] See an article by Mr J. E. Drower, in *The Times* of August 26th, 1926.

[2] *Report of the Royal Commission on the Coal Industry*, 1925, vol. I, pp. 156, 157.

has been estimated as 450 for union men[1]. Thus the total cost of labour has risen much more than the percentage increases in wages indicate.

As long as such relatively high costs continue, it is almost hopeless to expect that new smallholdings, or any other improvement in the capital equipment of agriculture that involves building, can prove remunerative. It is, also, difficult to get work done, at all events in remote districts—sometimes the supply of bricks fails; more often men are unobtainable. A large scheme, which gives a prospect of considerable profit, may be taken in hand, but the constant small repairs, needed to keep farm buildings in order, may wait for months before the local builder will attend to them. With the present carefully guarded entrances to the building trade, he cannot get men enough. There is much building work which can be learnt by an ordinary man in a few weeks. Is it not almost incredible that, with large numbers of unemployed, the nation tolerates trade union restrictions which prevent new recruits being put to urgently needed construction? Even the scheme of training apprentices, agreed on between the Government and the trade unions, is not being carried out, at all events in some places. The shortage of men will not get less while this continues.

Restrictions such as these involve an appalling economic waste. Together with the loss of work due to trade disputes, they do more than can easily be estimated to prevent the nation from recovering the losses of war, and starting afresh on a new career of prosperity and high wages. Lest I should be thought to exaggerate,

[1] See Mr J. E. Drower, *loc. cit.*

let me call in witness Mr Graham Wallas, the sometime
Fabian essayist, who cannot be suspected of bias against
trade unions. Mr Wallas writes[1]:

> In 1920...Parliament and the Cabinet, under the leader-
> ship of Dr Macnamara, challenged the building trade
> unions by a proposal to introduce rapidly trained demobil-
> ized soldiers into the trade. The Government were opposed
> from the first, openly by the men, and tacitly by the masters,
> and have now acknowledged their hopeless defeat. In this
> respect our position seems to be growing not better but
> worse....
>
> In the nineteenth century the railway industry, the iron
> and steel shipbuilding industry and the machine making
> industry owed their rapid growth to the fact that men were
> freely transferred to them from other industries that were
> decaying. Now our million and a half of unemployed persons
> include still unabsorbed soldiers from 1918, and unemployed
> engineers who are prevented from entering the building
> industry, and may soon include a fearful proportion of
> unemployed miners who will find every trade and every
> frontier closed against them.

Mobility of labour, both manual and directional, is
essential to healthy industry. One reason for the general
prosperity and high wages in the United States is the
readiness with which everyone turns his hand or brain
to any work that is wanted. In England it is becoming
more and more difficult for a man, however willing, to
do so, and for this economic viscosity trade unions are
chiefly responsible. Till we learn better ways, we cannot
hope to share American prosperity, or pay the high
wages which are there customary.

But let us return to the point which immediately concerns us. There seems little hope of getting the cost of
building into conformity with the prices of goods open

[1] *The Nation and the Athenaeum*, July 31st, 1926.

to free competition till the present shortage of houses is overtaken. The need is so pressing, and the importance of adequate housing to health and happiness so great, that the nation must now put up with being exploited and pay the bill. But a few years more should see the end of the present position. As soon as enough houses are available, and subsidies cease, the shortage of men and materials should cease also and prices come down. The repairs and maintenance of existing farm buildings and the equipment of new smallholdings, will then perhaps be possible at reasonable prices. Meanwhile, the critics of the landowner should remember that it is this high cost of building which is one of the chief obstacles to the maintenance and improvement of farm equipment, and that its cost would be the same whoever is responsible for the upkeep, whether an occupying owner or the State. Indeed, if the State were the owner, it is safe to say that the era of high costs would be prolonged.

CHAPTER IV

## SECURITY OF TENURE

IF perchance, not realizing the true causes of agricultural depression and its effect on methods of cultivation, we accept the view that English land is under-farmed, some of our land reformers, and among them the Liberal Land Committee, try to persuade us that it is largely due to the tenants' want of security of tenure. This idea has inspired most of the agricultural legislation of recent years. Nowadays the landowner cannot regain possession of a farm let on a yearly tenancy without paying heavy damages, unless he indicts his tenant for bad cultivation and the County Agricultural Committee endorses the charge. Even if a landowner be willing to incur the odium and trouble involved, it is unlikely that the Committee will make an order except in gross cases. The consequence is that "security" falls alike on the good farmers and on the bad.

The Green Book itself recognizes the failure of this policy. It says[1]:

"To the end of 1924, 531 applications from landlords or agents alleging bad farming had been received, and the County Committee had granted the landlord relief from paying compensation in 292 cases," while the lowest estimate of bad farming, 5 per cent. of the whole, means 20,000 to 40,000 holdings. The Book continues: "The recommendations made for increasing tenants' security of tenure by the Liberal Land Enquiry Committee of 1913 are in principle embodied in the present Agricultural Holdings Acts. The

[1] *The Land and the Nation*, pp. 446 and 448.

machinery, however, is not at all that recommended by the
L.E.C. The effect has to be acknowledged as disappointing.
Lord Bledisloe, in the remarkable address" he gave to the
British Association in 1922, declared that the Acts "have
really put the bad tenant on equal terms with the good, to
the great detriment of the industry. It seems to us," adds
the Report, "frankly impossible to give good farmers that
fuller security which they still require without stultifying
the community's claim to secure the proper cultivation of
the land."

Now this account of the effect of the present Acts is
sound, good sense. But, in spite of the acknowledged
failure of what has been done, the Green Book advises,
not a repeal of the Acts, but an intensification of the
policy contained in them. And, as another worse failure
is certain to follow, it proposes to upset the whole system
of British land tenure to give the policy a chance,
apparently in the hope that a new, more democratic
County Agricultural Authority will be willing to eject
bad farmers and secure an increase of food production.
But would this result follow? I doubt it. If the
County Agricultural Authorities resemble the present
Committees, they will be very slow to say that a farmer
is to lose his means of livelihood, and an increase in
security of tenure, as in the past, will intensify slackness
and inefficiency. If, on the other hand, the democratic
element in the Authority produces a change of policy,
it will be in the direction of the peasant outlook, of
leniency towards ineffective smallholders with a jealousy
of larger farms, even if efficient, and a determination to
take every opportunity of breaking them up. There is
ominous evidence of this in the descriptions given in the
Green Book of meetings held in villages to discuss the

proposed policy. The end would then be a return towards subsistence husbandry, and a shrinkage instead of an increase in external food production.

In *The Land and the Nation* much stress is laid on the idea that, without further security of tenure, good farmers cannot or will not improve their land for fear of having their rents raised or their farms taken from them and their expenditure and labour lost. Under the inspiring title of "Energy and Hope" a chapter is devoted to prophecies by farmers and others of the wonderful things that might be done if agriculturalists had the security the authors promise them. A Welsh farmer would be disappointed if, in ten years time, his "production had not increased by about 30 per cent. or 40 per cent.," though, modest man, he cautions us that "the increase after that would be at a slower rate." The assumption all through is that in present conditions a farmer may lose the benefit of any improvements he makes, and that agriculture therefore suffers from the tenants' want of security.

To answer this contention, one has only to turn to another part of the Green Book (pp. 101–3) in which arguments against multiplying ordinary freehold ownership are needed.

A very striking feature of the comments made at our Enquiry Campaign meetings was insistence on the fact that land owned by its occupier is frequently the worst farmed land in the district....Of a certain district where owner-occupancy predominates our investigator reports as follows: "All the soils in this district are very short of phosphates and most of them of lime; where these natural deficiencies have been rectified, the farms look very well...there are considerable communities in this district where the occupiers are devoid of capital, experience or knowledge of modern farming, with no education, and unaccustomed to contact with the outside world.... Their standard of living is considerably below that of the usual farm worker both as regards housing, clothing, food and recreation. The land is mostly owned by the occupiers and is a good example of what many districts will descend to in the absence of an efficient land-owning class, if a determined effort is not made by the State to introduce a properly organized system of land tenure."

The authors seem to overlook the fact that, since occupying owners have the most absolute security, this section of the book refutes that which claims that tenants cannot farm well owing to want of security.

Moreover, it appears to be entirely ignored that the danger of a farmer losing the benefit of his improvement —if indeed it had ever any basis in fact, save in the few hard cases which make bad law—is entirely met by the present Agricultural Holdings Acts. Firstly, a tenant has complete freedom of cropping. He can cultivate his land as he pleases. Secondly, the value of his improvements are secured to him. Not only can a tenant claim from one to two years' rent as compensation for disturbance if his landlord gives him notice, but, however he leaves his holding, whether voluntarily or involuntarily, he is paid for improvements which he has made, with or without the consent of the landowner, for drainage, for chalking, liming or manuring the land, for laying down temporary pasture, for repairs to necessary buildings, and for increased value of the land due to a specially high standard of cultivation[1]. Indeed, he has only to get his landlord's consent for any other kind of improvement in order to secure compensation for that also when

[1] Agricultural Holdings Act 1908; Agriculture Acts 1920, 1921.

he leaves. Even before the present stringent laws, though a few cases of hardship may have occurred, the advantage to a landowner of keeping a good tenant, to say nothing of the friendly feeling which usually existed, was enough to provide practically perpetual tenancy.

To put it plainly, this outcry about "security of tenure" is very largely nonsense. Landlords are generally quite willing to grant leases; it is the tenant who demands a yearly agreement. He knows he is most unlikely to be turned out, and he retains his own freedom. On a yearly agreement it is heads he wins and tails his landlord loses. Anyhow, as stated above, under the present law he gets compensation on leaving for improvements he has made, and, if his landlord gives him notice, compensation for disturbance. He has too much security—not too little—and, good fellow though he usually is, he is sometimes inclined to take things easy in consequence.

I have no sorrow for the farmer's insecurity of tenure though all the Liberals in England weep sympathetic tears: as regards his landlord, he can look after himself. There is nothing to prevent a tenant from farming in the best way the economic state of the industry allows, and carrying out any improvements that he thinks will repay their cost. If he is not doing so, it is his own fault.

Indeed, as Sir Henry Rew has pointed out[1], while the present system of land tenure bears a misleading external likeness to that which held in the nineteenth century, it has in reality been quite changed. The landowner's power of securing a definite method or general high standard of farming has been destroyed by suc-

[1] British Association, 1926, Section F.

cessive Acts of Parliament. Nothing effective has been put in its place, and a tenant can farm well or ill as he chooses, practically free from all control. His landlord's influence can now only be exerted by persuasion and example, and critics, when they complain that landowners have "ceased to lead," should remember that the owner's power of insisting on good husbandry has been taken away. His modern part is more difficult to play, and can only be made effective by a combination of knowledge, tact and determination.

As an example of what land reformers wish to put in place of the present system of tenure, let us turn to the scheme of tenancy proposed by the Green Book. The land of each county is to be managed by a County Agricultural Authority—a democratic committee, about whose constitution and powers we shall have something to say below. Subject to this committee, or its officers, being satisfied that his cultivation is up to their standard of requirements, the "cultivating tenant" is to be given absolute security at a fixed rent as long as he continues to farm well, with power of bequest to a son or other near relative able to carry on the farm. There is no provision for the friendly remissions of rent which characterize present dealings of landlord and tenant in bad times. Nevertheless, as Colonel Tomkinson pointed out[1], by leaving or threatening to leave his holding, the tenant can oblige the County Authority to exercise its power of re-assessing rent on a vacancy, and thus rents will be forced down in times of agricultural depression. On the other hand, when times improve, rents will remain fixed, and every farmer will find a son or other

[1] *The Times*, November 30th, 1925.

relative to take on property rented below its economic
value. From the national point of view the whole
financial scheme is radically unsound. And this defect
is not one which can be cured by modifying details. It
is inherent in any scheme which gives a tenant security
of tenure while leaving him free to throw up his holding.
It can be avoided only by occupying ownership on the
one hand, or by a return to ordinary, honest, straight-
forward terms of tenancy on the other.

It will be a bad bargain for the State, and I think that
very few farmers, in whose interest the scheme is pro-
pounded, would accept it in exchange for the present
system of land tenure. They would gain, it is true, at
the expense of the nation in times of rising prices, when
they could claim security of tenure at fixed rents, but
let us examine the other conditions under which they
would hold their farms.

The rent is to be a "fair net rent" calculated on the
landowner's present net receipts corrected for any
addition to the minimum wage of the agricultural
labourer, effected in some unexplained way, at the time
of the change. But the tenant will be responsible for
his own repairs; he will be exposed to any further
addition to wages with none of the protection now
afforded by a friendly reduction of rent if times become
too hard; he will be subject to the uncertain requirements about good cultivation of an inefficient County
Agricultural Officer, or the constant pressure of a better
man anxious to demonstrate his efficiency. It is safe to
assume that the average farmer, though he may possibly
grumble at his landlord to please an itinerant investigator, would rather continue under the present system

than fly to so many evils that he knows not of. What do
the Liberal Land Committee think would be the result
of a ballot of farmers? Do they think that more than one
in ten would support their scheme?

In conclusion, we must point out that Mr Orwin and
Colonel Peel propose to continue ordinary tenancy under
County and District Land Agents responsible to Whitehall. Even the Labour pamphlet avoids the pitfalls of
the Liberal scheme of "cultivating tenancy." It says:
"Present agreements and customs and conditions of
tenancy will continue under public ownership until modifications are required to meet new circumstances." It
adumbrates a visionary plan for "forms of collective
or co-operative farming" in the distant future, but it
accepts our present system of tenancy for the immediate
purposes of public ownership of agricultural land.

The arguments used by some Liberals in favour of
additional security of tenure are also adduced among
other reasons by those who advocate a large extension
of occupying ownership. The advantages of possession
are largely those of security. Improvement of the land
is for the owner the improvement of his family property,
and it is natural to expect that more improvement will
be made. Yet, in so far as improvement can be measured
by increased production, there is little statistical evidence to bear out our *prima facie* expectation.

The Agricultural Tribunal[1] finds that in the United
States there is "a persistent increase in tenant farming."
This seems to be regarded by some people in the States
as "alarming," but it must have some real economic
advantages for it to persist. A comparison of crop yields

[1] *Report*, p. 343.

per acre on owner and tenant farms "shows that the productivity is slightly less on tenant farms." But a more recent enquiry by Sir Henry Rew[1] led to the conclusion that there was no significant difference in America, and that other factors were much more important than the system of tenure in determining productivity.

The case of Denmark is often brought forward to support the benefits of ownership. But we must not forget that in Denmark one-third of the population is actively employed in agriculture, which is the only large industry in the country, none other competing seriously with it for brains or labour. Denmark has solved the problem of keeping on the land a large number of prosperous and contented smallholders, who are prepared to work harder and for longer hours than is now customary with English farm labourers. Danish agriculture clearly impresses those who know it well. They never tire of extolling it as an example to our farmers. Yet, as explained in Chapter II, Macgregor's investigation shows that the corrected values for the agricultural productivity of different European countries brings out Denmark on a par with Great Britain. There is a higher proportion of arable land, and so the total output of food per acre is greater, but, as far as the yield of crops corrected for the proportionate area is concerned, an acre of plough land in Denmark gives no more than an acre of plough land in England[2].

Secondly, although Danish tenure is called "ownership," it is qualified by custom which has acquired the

[1] British Association, 1926, Section F.
[2] See an extract from Macgregor's paper, p. 33, above.

force of law in so far as a farm cannot be subdivided or joined to other farms[1]. Moreover, the cottage holdings created by a law of 1919 are held subject to further restrictions. The "owner" is not called on to buy the holding, but only to pay an annual charge based on the valuation, which is liable to re-assessment every five years. The State has a prior right of resumption if it is wished to transfer the land to anyone not the heir; the land must be used for agricultural purposes only, the necessary equipment must be maintained and no injury done to the land; the holding cannot be let.

The name "ownership" is retained in deference to the dislike of the Danish peasant for the idea of tenancy, but these qualifications seem to justify the authors of the Green Book in their contention that the State small-holdings in Denmark are in much the same position as they propose for England under the name of "culti-vating tenure." At all events, Denmark seems an un-certain example to stress in favour of the creation of large numbers of small unrestricted freeholds with the idea of increasing the agricultural output from our land.

If the Danish peasant prefers ownership, there is no doubt that the English farmer prefers tenancy. Orwin and Peel quote from the *Report* of the Haversham Committee of 1912:

The evidence we have heard makes it quite clear that tenants do not desire to purchase their farms except as an alternative to leaving altogether (Sec. 76).

Of farmers who gave evidence before us, only three advo-cated purchase save as an absolute necessity. One land agent stated that putting himself in a tenant's position, nothing would induce him to buy; another was of opinion

[1] *Agricultural Tribunal, Final Report*, pp. 258–64.

that no tenant desired to purchase except under compulsion (Sec. 77).

In the same way the experience of all County Councils since the Small Holdings and Allotments Act, 1908, came into force shows that the great desire of the applicants is to rent land; in fact, only 2 per cent. desired to acquire it (Sec. 81).

I think that all those who come much in contact with farmers will agree that, since the lesson of the last five years, they are even less willing to buy than they were in 1912.

The diffusion of property is desirable, and large numbers of small landowners would be a sound and stable element in the nation. But it is difficult to force a policy against the wishes of those for whose benefit it is proposed. I believe the "magic of property" works as well in livestock and crops as in the freehold of land, and I think the benefits which may reasonably be expected can mostly be secured by an extension of the present system of County Council smallholdings and cottage holdings, on the lines of the Government Bill now before Parliament, which also provides for any who may wish to purchase their farms. The statutory limit of size to a "smallholding" should be raised or removed.

And there is another point. The aesthetic consideration of the beauty of "England's green and pleasant land" may seem out of place in that part of a book which is concerned with agriculture. Yet that beauty has a very real value—moral, educational and even financial. The small owner has rarely much taste or any vivid sense of beauty. Too often his first act on buying a farm is to cut down every tree for the few shillings it will bring and the few square yards of land it overshadows. If he builds anew or repairs old structures, his style is usually lamentable, and small freeholds, with hideous houses and septic-looking eruptions of sheds made of second-hand wood and corrugated iron, often become an eye-sore to the landscape. One has only to watch for a few years the changes that go on in any English countryside, to see many instances of the aesthetic and historic crimes committed by the small owner.

But will national ownership be any better? The Green Book picks out for reprobation a list of advertisements of small estates and farms such as:

Gentleman's Queen Anne residence with...125 acres.
Gentleman's rich dairy farm; 180 acres; intersected trout stream; fine Tudor residence.

The authors' comment on these advertisements is as follows:

Greater confidence could be felt in the present condition and the future prospects of English agriculture if typical advertisements of farms to let had a good deal more to say about fertility, pig-sties and cattle-sheds, and a good deal less about Tudor residences and trout streams[1].

Such remarks fill one with amazement. Surely it is better that the possession of our priceless treasures of domestic architecture should pass to those who appreciate and can afford to care for them, than to those whose sole ideas are pig-sties and cattle-sheds! Moreover, the educated man who appreciates the Tudor house is likely with experience to make the better farmer— none the less because he is a good sportsman and loves a trout stream.

[1] *The Land and the Nation*, p. 266.

The profits of farming are small, but many a man with other resources chooses a farmer's life for better reasons. The amenities which attract him are not lightly to be destroyed. They bring satisfaction to the individual, and through him economic and social benefit to the rural community.

There is no safeguard from vandalism in schemes in which the "cultivating tenant" has to do his own building and repairs, and the "democratic County Agricultural Authority" may be as blind to beauty as the small owner. Some such provision as that of Orwin and Peel might, it is true, combine aesthetic with agricultural control. But the real lesson is that wisdom lies in the preservation of the present race of landowners, where that is possible, and an extension of the amount of land held by County Councils, rather than in either revolutionary changes on the one hand or the multiplication of small freeholds on the other.

The farmer's troubles have nothing to do with want of security of tenure. As the Farmers' Union consistently and continually points out, they are caused by unremunerative prices. To this result all our enquiry has led up, and, leaving subsidiary points, we must now pass to an investigation of the real factors in agricultural prosperity and adversity.

### Chapter V

## AGRICULTURAL DEPRESSION

It will be seen that most of the circumstances on which are founded the criticisms aimed at British landowners and farmers are really due to the recent depression in agriculture—to high costs and unremunerative prices. For any discussion on agricultural policy, an understanding of the causes of the periodical general depressions from which agriculture suffers is clearly necessary. Yet that understanding seems very rarely to be part of the equipment of the disputants, who, in accordance with their individual prepossessions, talk about foreign competition and free trade, insecurity of tenure and "landlordism," or the breakdown of capitalism and the socialization of industry—all equally beside the point.

The chief cause of agricultural prosperity or adversity is a combination of two factors, the recurrent rise and fall in the general level of prices, and the economic lag between expenditure and receipts in farming operations. We will discuss the latter of these factors first[1].

The growing of a crop is a slow process. A farmer has to prepare the soil before the crop is sown, and part of this work is done while the crops of former years are in the ground. Hence much of the cost of production is incurred many months or even a year or two before sale. Now, by properly weighting the different items of cost, it is possible to calculate the average time before a crop

---

[1] See *Journal of the Royal Agricultural Society*, vol. LXXXV, 1924; *Economic Journal*, December 1925.

is sold that the cost of growing it is incurred. The
financial results will be the same as though all costs
were paid at that time. This lag between expenditure
and receipts, with the normal methods of agriculture in
the South of England, is found to vary from about
7 months for a grass land dairy farm to about 14 months
for a typical arable sheep and corn farm.

Now the lag, which is much greater than in most
other industries, has a very important influence on
agricultural economics. If prices are rising, a farmer

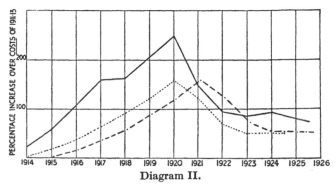

Diagram II.

incurs his costs at a lower level, and, when he sells, he
makes a fortuitous profit. On the other hand, if prices
are falling, he incurs his costs at a high level and sells
when prices are lower, sometimes for an amount which
is less than his costs of production. Hence arises the
great importance of variations in price in agricultural
economics.

The effect of economic lag is best illustrated graphically. In Diagram II, the highest curve shows the percentage increase in price of the produce for a typical

light land sheep and corn farm from 1914 to 1926, over the average price of the same produce from 1911 to 1913.

The dotted curve gives similarly the percentage increase in the total costs, counting in labour, manures, feeding stuffs, implements, rent, and all other outgoings. Since from 1911 to 1913 arable farming was making a modest profit, the higher or receipts curve is started enough above the costs curve to indicate a profit of 20 per cent. on the turnover, perhaps 10 per cent. on the capital employed[1]. But since, on an average, the expenditure on an arable crop was incurred about 14 months before it was sold, the dotted costs curve in Diagram II has to be shifted to the right by a distance representing 14 months, to bring the receipts from and expenditure on a crop into the same vertical line. This is done in the third or chain curve of the diagram. In a similar figure for a grass dairy farm the shift would be one of only 7 months.

Now, when the curve giving the increase in the price of produce lies above that giving the increase in cost, it indicates that an average farm of the type described might be expected to make a profit, the receipts being more than the costs of production by an amount measured by the area enclosed between the two curves[2]. It is seen that, when prices are rising, this area is increased

[1] The profits have to cover wages of management earned by the farmer himself and any interest on borrowed capital or bank loans.

[2] It should be noted that the diagram refers not to an actual farm, but to an average or typical farm of the kind named. Also the area between the two curves measures not the total profit or loss, which involves a capital valuation, and not the balance of yearly income and expenditure, but the profit or loss on the year's receipts shown by a properly kept costings account.

by the shift of the costs curve, so that profits were much larger from 1914 to 1920 owing to the economic lag. On the other hand, from 1921 to 1923, when the cost curve lies above that indicating receipts, the loss is very greatly swollen by means of the lag, so much so that most crops brought in less than the cost of growing them. The general effect of the lag, then, is to increase the risks of farming.

On a grass dairy farm, the lag is only about 7 months, half that on the arable farm. Hence we see the sound economic reason which underlies the general opinion that grass land is safer than arable land. We see also why financial results are so overwhelmingly affected while a change in price is going on. A farmer, especially an arable farmer, is almost sure to make profits when prices are rising and suffer losses when they are falling.

The general level of agricultural values, measured in the amount of other goods and services that agricultural produce will purchase, depends primarily on the cost of growing crops and rearing stock compared with the cost of supplying those other goods and services. These values can be expressed in terms of money as prices.

Hitherto, much of the world's supply of food has come from peasant proprietors in old countries, with a comparatively low standard of life. These peasant farmers drive down the profits and wages of those with whom they compete, both here and in new lands overseas. The drift to the towns is now spreading to all nations, and since the war even peasant proprietors are learning to leave the land. Moreover, the population of the world is still rising, and the amount of new land is limited. Both these changes will tend to cause a check to the output of food and thus a slow but permanent rise in agricultural values and therefore in the general permanent level of agricultural prices compared with others, in the not very distant future.

On the general price level thus determined, variations are superposed. Variations in the price of agricultural produce are of two kinds: firstly, a variation confined to one commodity, *e.g.* wheat or potatoes, and due chiefly to plenty or scarcity over the area supplying the market; secondly, a variation in the general level of prices of all commodities, in which agricultural produce shares equally with other goods. The first kind of variation will be dealt with in a later chapter. It produces very great uncertainty and sometimes heavy loss—when, for instance, pigs have been reared in too great numbers and fetch less than their cost, or a good crop of fruit is unsaleable owing to a glut in the market. The cure for these troubles is probably co-operation to secure orderly marketing.

But such variations do not cause the general depression which overtakes the industry from time to time. From 1873 to 1896 and from 1920 to 1923, the prices of all agricultural produce fell disastrously, and prices of other commodities shared in the fall. This general fall in prices looked at from the other side is, of course, an increase in the amount of goods or services a pound or a shilling will buy—a rise in the purchasing power of money, that is, a rise in its value.

This second kind of variation can be further subdivided into short-term fluctuation, following the cycles of good and bad trade, which recur every few years, and long-term change—a slow drift of prices underlying the wave-like fluctuations of the trade cycle—such a change, for instance, as the long fall from 1878 to 1896, in the years of the great agricultural depression.

To illustrate the short-term fluctuations, let us look at the effects of the war. The following table gives in parallel columns the rise and fall in the average price of raw materials other than food, and of the index number giving the average price of agricultural produce, as published by the Ministry of Agriculture.

TABLE II

*Percentage Increase in Price over the Average of* 1911–13

| Year | Agricultural produce | Raw materials | Year | Agricultural produce | Raw materials |
|------|------|------|------|------|------|
| 1914 | 1 | 0 | 1920 | 192 | 194 |
| 1915 | 27 | 23 | 1921 | 119 | 111 |
| 1916 | 60 | 59 | 1922 | 69 | 74 |
| 1917 | 101 | 103 | 1923 | 57 | 64 |
| 1918 | 132 | 133 | 1924 | 61 | 72 |
| 1919 | 158 | 152 | 1925 | 59 | 61 |

It will be seen at once how closely the two sets of figures agree. Here we have a change which has to do, not with factors which concern one commodity only, but with some deep-seated cause affecting all things alike.

Similarly, with the long-term drift of prices—subject to individual differences due to special causes, the average prices of all commodities move together. During the period beginning with the years 1871–5 and ending with 1894–8 the average price level of all wholesale goods fell by 40 per cent. The percentage falls in the chief kinds of agricultural produce were wheat 51 per cent., barley 39, beef 29, mutton 25, pork 33, oats 38. The index number of agricultural produce was not yet in being, but, from the arithmetic mean of these figures, we get an average agricultural fall of about 36 per cent., roughly equal to the 40 per cent. fall in all commodities.

What, then, are these irresistible causes which bring about the ebb and flow of the industrial and agricultural tide, on which some manufacturers and nearly all farmers alike drift almost helplessly to prosperity or adversity? The problem is best approached by a consideration of the slow, long-term changes, such as is found in books on Economic History like W. T. Layton's *Introduction to the Study of Prices*[1].

For thirty years after the close of the Napoleonic Wars prices were falling and agriculture was depressed. Ignoring the violent change of the earlier years, Layton gives the fall in the average level of wholesale prices from 1823 to 1848 as 25 per cent. It will be noticed that during this period the protection afforded by the Corn Laws did not prevent acute agricultural depression. From 1848 to 1873, prices slowly rose by 20 per cent. The prophecies of disaster which were made on the repeal of the Corn Laws in 1846 proved baseless. In spite of free trade and cheaper transport, farmers did well, the area of land under the plough increased and reached its maximum, and British agriculture gained its highest point of success and renown.

About 1874 the tide turned. Prices began to fall, and the wet harvest of 1879 ushered in the long agricultural depression, which lasted with increasing gloom till the

[1] Macmillan and Co., new edition, 1922.

closing years of the nineteenth century. The lowest level
of price was reached about 1896. Thereafter a recovery
began, and the few years before 1914 showed modest
agricultural prosperity.

It is clear that farming is successful in times when the
average level of wholesale prices is rising, and has to
face hard times when that level is falling. As in the
shorter cycle of recent war years, agricultural prices tend
on the average to rise and fall with the index number
of other wholesale commodities. This being so, in our
search for causes, we must look, not to factors that concern agriculture alone, but to those that lie deeper, and
enter into the prices of all goods and services.

What is price? It is the number of units of currency
which buyer and seller agree on for a unit of the commodity bought and sold—the number of pounds taken
for a ton of copper, or the number of pence given for
a pound of tea. By law, twelve pence are equivalent to
one shilling, twenty shillings to one pound, and, in whatever way money was transferred, whether by notes or
cheque, one pound, from 1820 to 1914, could be exchanged for a coin containing 113 grains of fine gold.
Thus the value of our currency depended on that of
gold. Experience showed that, with a certain reserve
of gold in the vaults of the banks, it was safe to issue
notes and other instruments of credit to a much larger
amount, for those who held them did not all claim gold
at once. But, as long as the notes were convertible at
will, the value of pounds, shillings and pence depended
on the value of gold. A purchase is, then, a disguised
barter, a ton of copper or a pound of tea being exchanged
for an order printed on paper or silver, and guaranteed

by the Government or the Bank of England, for a certain number of grains of gold.

Now gold, like other metals, is a commodity, with new supplies only to be dug from mines, and in demand for currency, plate and jewellery. Its use can be economized—the thinnest film of gold-leaf is enough to beautify a picture-frame, and a few score millions in the vaults of the Bank of England are enough to support our vast superstructure of currency and credit. But the fact remains that we can afford to gild more frames, and can safely create more currency and credit, when new mines are discovered and gold becomes more plentiful.

But, before we can fully comprehend the causes which fix the general level of prices, another point must be made. We do not, unless indeed we are misers, acquire pounds or shillings to hoard them. We barter them again for bread, or theatre tickets, or stocks and shares. Hence a given amount of currency will do more work, and therefore be more effective, the oftener it is exchanged, and, on the supply side of our equation, we must multiply the total of currency and credit (let us call it $c$) by ($v$) its velocity of circulation, that is, by the number of times it is turned over in a month or a year.

The other side of the equation must represent the demand for currency and credit—the amount of business ($m$) transacted in the same time multiplied by the average price ($p$) at which it is done. Thus

(currency and credit) × (velocity of circulation)
= (amount of business) × (average price),

$$cv = mp$$

or
$$p = \frac{cv}{m}.$$

Hence we see that the average level of prices will tend to rise if $c$, the amount of currency and credit (which ultimately depends on the supply of gold) or the velocity of circulation $(v)$ increases, and will tend to fall if there is an increase in $m$, the amount of business to be carried on. If the currency and credit, or rather $cv$, increase at the same rate as $m$, the business to be done, the average price level remains constant.

Now, while methods of banking and general business habits remain unchanged, the velocity of circulation $(v)$ will be roughly constant, and the price level will chiefly depend on the relation between the supply of gold and the demand for it produced by the growth in business. And this fact explains the broad changes in price during the last century.

From 1820 to 1848, the closing of many South American mines and the vast expansion in industry and commerce led to a shortage of gold and currency. Gold became scarce, and therefore dearer in terms of commodities. That is, more goods had to be given to secure a pound or a shilling, prices fell, and farmers suffered.

Then, at the end of the forties, new mines were opened in Australia and California, the world's annual production of gold increased six or sevenfold in a few years, and gradually the whole stock of gold in circulation expanded. Gold became more plentiful and therefore cheaper—more of it had to be given for a quarter of wheat or a ton of iron. Prices first became steady and then began to rise. Very soon most farmers found that their receipts exceeded their expenses, the herald of good times.

The world's annual output of gold remained about the same for forty years, but in the early seventies the demand was increased by the adoption of a gold standard for money by Germany, the United States and the Latin Union, and the consequent call for more gold reserves. Demand in fact once more outran supply. The value of gold rose, that is, the value of other things measured in gold, *i.e.* prices, fell. People with nominally fixed incomes gained, for their incomes would buy more. Some industries that could take advantage of new inventions, or otherwise lower their costs, gained by cheapness and the expansion in trade that cheapness brings. But agriculture, with its five thousand years of experience, is benefited more slowly by the growth of knowledge, and the long economic lag of farm crops makes it impossible to adjust costs in time to meet the losses of a falling market. Hence the years from 1875 to 1896 saw unrelieved agricultural depression.

It is usual to attribute the troubles of that time to cheap transport and foreign competition. But the fall in price was not confined to agricultural produce; the average price of all wholesale commodities fell by 40 per cent. Even in the case of wheat, the fall was only 50 per cent., of which 40 per cent., as we now see, must be assigned to causes which affected goods other than agricultural. Most other kinds of agricultural produce fell by less than the average, that is, actually increased in real value when the change in the value of money is allowed for. It is clear that the preponderating cause of depression was a shortage of gold, which drove down the general price level. The depression was felt more in agriculture than in other industries, partly because of its long economic lag, and partly because its costs of production could not be much diminished by the application of power and of new inventions.

It must be noted that the agricultural depression was not confined to England. It was equally severe in countries blamed by the British farmer for their competition, with the significant exceptions of India and Argentina, where alone in those years farming flourished and land rose in value. Now the coinage of India was based on silver, and Argentina used only paper money, so that neither of these countries was affected by the shortage in gold.

This point is so important that it is well to copy the paragraphs dealing with it in the *Report* of the Committee on Stabilization of Agricultural Prices[1]. Referring to the great depression of 1874 to 1896 and the Royal Commission, which investigated it and issued a first Report in 1894, the Committee state:

The fall in prices was in turn attributed to foreign competition consequent upon the development of the new countries and the cheapening of the means of transport. This has in fact become the almost universally accepted interpretation of the great decline in British agriculture between 1874 and 1896. The progress of invention, the ever-increasing area under corn in the new countries, the rapid development of rail and ocean transport, all contributed to increasing the foreigner's competitive advantage in the markets of Great Britain, and to this was attributed the great fall in corn prices and consequent decline in the arable area. Such being the general belief, it is not unnatural to find that the remedy loudly demanded in many quarters, and almost invariably advocated by witnesses before the Royal Commission, was a return to a system of protective tariffs.

[1] *Report of the Committee on the Stabilization of Agricultural Prices*, Ministry of Agriculture, Economic Series, No. 2, 1925, p. 83.

One point of outstanding importance is the fact that agriculture passed from a period of great prosperity to depression with comparative suddenness during or about the year 1874. But the progress of invention, the development of the new countries, the cheapening of means of transport, etc., had gone on steadily for 20 years before 1874 and continued comparatively steadily throughout the period of depression. There was no sudden change in these conditions which coincided with the change in fortune of British agriculture in 1874. To this point attention was called by Sir Robert Giffen in his memorandum submitted to the Royal Commission in 1896. "The phenomenon to be explained," he says, "is why the causes referred to, produced no fall in prices before 1873, but were even consistent with a rise, and why they produced a fall after 1873."

What is the explanation? The plain fact is that at this time, just as in the crisis of 1920 to 1923, agriculture was struggling against an economic environment of which agriculturalists were almost entirely ignorant. Influences were at work which had brought about a depression far more widespread than the British farmer was aware of; even in the great new countries, whose competition was so bitterly complained of, a severe fall in prices was also being experienced.

A supplementary Report attached to the Final Report of the Royal Commission published in 1897 and signed by about half the members of the Commission pursued the enquiry a stage further. It examined the conditions of agriculture in other countries during the period of the British depression, particularly in the great protected countries, Germany, France, the United States, Australia and New Zealand.

From all these countries the story which was told was much the same, the same symptoms of economic distress were manifest. In some places conditions were serious, even critical, and everywhere diminished profits or widespread losses were reported. In France the price of wheat, in spite of drastic increase in the duty on imported wheat, declined by 43 per cent. between 1867 and 1894. In Germany the prices of wheat and rye fell to a point "admittedly below the cost of production" and the losses on the crops in the years 1893 and 1894, after allowing for the cost of production, were estimated at no less than £13,000,000. In Russia, the United States, Denmark, the Netherlands, Australia and New Zealand, much the same conditions prevailed. "All these countries," the report stated, "complained in a greater or less degree of agricultural depression, which in each case is attributed to precisely the same cause, viz. the fall in price of agricultural produce." In no case had protection prevented a depression in agriculture.

But to this general and widespread depression there were two important exceptions, namely India and the Argentine. From India it was reported that agriculture was flourishing, the land under cultivation had extended, the number of stock increased, exports had risen, the rental and revenue from the land grown, and prices of agricultural produce had been maintained. Similar reports were received from the Argentine. What was the reason? Why were these two countries an exception to the apparently almost universal rule? Were there any circumstances in common amongst the countries which had suffered an agricultural depression which were not shared by the other two? The answer to these questions was a very simple one. The common factor shared by these [depressed] countries was that their currencies were on a gold basis. [On the other hand] India had a silver and the Argentine a paper currency.

Now the monetary events which had taken place during the period under examination had an extremely important bearing on the question. Not only did the depressed countries have in common a gold basis for their currency, but they were affected by monetary conditions in an even more striking manner. Prior to 1873 the mints of the United States, France, Switzerland, Italy, Belgium and Greece were open to the free coinage of silver as well as gold. In other words these countries had bi-metallic currencies; Germany had a silver currency. About that time the position was entirely changed, however, by the passing of measures hostile to silver in European countries. Germany changed from a silver to a gold currency, following upon which all the above countries closed their mints to the free coinage of silver. Italy, the United States of America, Austria-Hungary, Russia and

Japan, were establishing a gold standard and buying gold. The result was a greatly increased demand for gold; the supply became inadequate, and the shortage was accentuated by a falling off in production at the mines. Consequently, the prices of commodities in terms of gold continuously fell in all the gold countries. India, on the other hand, remained on a silver basis, and Indian prices remained steady. It was a monetary revolution of almost world-wide character, and it was an event which *precisely* coincided with the sudden change from prosperity to depression in agriculture, both in Great Britain and abroad. The supply of purchasing power within these countries had been seriously contracted.

But in 1886 gold was discovered in the Transvaal, and in ten years the new supplies had begun to influence the total amount in circulation. From 1896 onwards, prices began to rise, and in a few years agriculture showed signs of revival. This revival, indeed, had been foretold on the basis of the rising production of gold by Sir Robert Giffen in his evidence before the Royal Commission in 1896. From 1904 to 1914 farming was in a healthy state, with all the symptoms of returning prosperity.

At the outbreak of the war in 1914 the gold standard was abandoned, while notes, in effect inconvertible, were issued in increasing numbers, and bank loans and deposits enormously expanded, to meet the huge demand for currency and credit to carry the hectic business of war trade and finance. Money, more in quantity and circulating faster, fell in value rapidly, and, though the fixing of prices put a limit to the rise in the nominal value of agricultural produce, on the average it rose as much as did other commodities. The post-war boom in trade still further swelled bank loans and deposits, and with them currency and credit. This in turn still further raised prices and reacted on trade. The process became cumulative—rising prices swelled credit, and swelling credit raised prices. The country was perhaps in danger of losing control of the currency and seeing it inflate until it became valueless, as it did in Russia and Germany. But in 1919, at the instance of a Treasury Committee presided over by Lord Cunliffe, it was determined to stop further inflation with the idea of a progressive return to the gold standard and to the old definition of the pound. Bank Rate was raised, and credit contracted. In 1920 prices broke in almost all countries of the world. In Japan and Great Britain the fall began in April; in France and Italy in May; in the United States, Germany, India and Canada in June; in the Netherlands in August; and in Australia in September. The action taken on the recommendation of the Cunliffe Committee may have been no more than a contributory cause. But however the fall began, once started it became cumulative—falling prices contracted credit, and contracting credit lowered prices. Once again the farmer, owing to the nature of his business, suffered more than most men during the fall. In the United States his position was even worse than in England, partly owing to the fact that there ownership rather than tenancy predominates, and much of the land is mortgaged[1].

By 1922 the drop had ceased in America and by 1923 in England. It looked as though a steady level of prices had been reached, giving a chance for a recovery in trade. Nominal wages, even in the most sheltered industries, had fallen, though less than in others, and costs had roughly adjusted themselves to receipts.

[1] *The Agricultural Crisis*, Washington, 1921, p. 205.

Agriculture regained its equilibrium sooner than some other unsheltered industries, chiefly for four reasons: (1) farmers had reserves owing to the profits of recent years, and had not invested much of them in undue expansion of their own business; (2) agricultural wages fell more than some others; (3) rents, always lower in England than elsewhere, had been raised much less than other costs, so that real rents were actually lower, and some of even this modest rise in nominal rents was remitted by many owners; (4) the burden of rates was lightened by the Act of 1924.

In 1923 and 1924, then, we appeared to have reached a position of stability. People began to breathe freely, and look for a slowly built revival of prosperity. The pound sterling had risen from its lowest point of 3·5 dollars, and stood at about 4·68 dollars, instead of the pre-war 4·86⅔. It had remained approximately constant for some months. Wages had fallen from their highest figures, and, though complaints were heard that men were earning less money than three years before, it was generally recognized that, as each shilling would buy more, they were nearly as well off. But difficulty yet remained. Wages had not fallen equally. In unsheltered trades, working for export, like coal, iron and engineering, or exposed to world competition, like corn farming, the prices of the products compared with the prices in foreign currencies had perforce fallen with each rise in the value of the pound. Each pound being worth more, fewer pounds could be obtained for a ton of coal, or iron, or wheat. Hence in unsheltered trades wages had to be lowered to keep going. But, in sheltered industries like railways, building and the distributive trades, a natural monopoly kept up profits and wages, and the fall had been much less. The cost of the services these industries rendered to others was kept high, and this added to the difficulties of the unsheltered trades. The economic machine worked, but was strained almost to breaking point by the discrepancy, coming, as it did, on the top of a lessened world-demand for both coal and iron.

The adjustments needed to meet a rise in general prices are much easier to make than those required in a fall, since they consist in raising wages to meet increased costs. A fall in prices involves a corresponding fall in nominal wages, and although, owing to the smaller cost of living, real wages may be unaltered, it wears an appearance of an attack on the workman's standard of life. Trade unions naturally oppose it. They can do so more successfully in the sheltered industries, and thus the postman, the transport worker and the bricklayer maintain their high wages at the cost of the exporting coal-miner, the skilled engineer and the farm labourer. In 1924 these discrepancies had appeared, though a position of apparent equilibrium had been reached.

Then, in the interests of national credit, and of London's international trade and finance, it was determined that the time had come to return to the gold standard, and to return to it, not at the then existing value of the pound, but at the old pre-war parity of exchange. The impending act and its accomplishment forced up the gold value of the pound by the residual 5 per cent. needed to bring it to its old par value. Each pound was suddenly worth more gold, or wheat or coal, and the change of course lowered the price measured in sterling of all world commodities. This last fall, coming when wages and other costs in the sheltered industries had been steady for some time at figures which had come to be accepted as normal, again injured the unsheltered industries, already depressed, and proved too heavy a burden for some of them to bear. The coal trade was faced with the alternatives of ruin or lower wages, iron and engineering saw their slight signs of revival vanish, and agriculture was plunged once more into gloom. The actual fall in price was small, but it was forced artificially in a time of temporary stability, and therefore cracked the economic machine, already strained, and finally threw it out of gear. In old conditions a rise in sterling would have resulted in a fall in all prices and in all nominal wages. But with powerful trade unions entrenched in the sheltered industries, costs in them cannot now be readjusted automatically. The unsheltered trades had thus to face not only an immediate decrease in the price of their products, but also the load of unaltered costs for transport, local services and distribution. We see, in fact, that the result of a rise in sterling is actually to subsidize the sheltered trades, both as regards profits and wages, at the expense of the unsheltered, and this subsidy continues till costs, including profits and wages, are lowered to the same amount in the sheltered trades likewise, so that their goods and services can be sold to the unsheltered trades at rates appropriate to the lower general price level. This concealed subsidy might be used as an argument in favour of open protection or subsidy, temporarily at any rate, for agriculture and other unsheltered industries. But the economic complication and the vested interests thus created weigh heavily against such a proposal. We have considered it more fully on p. 27. To subsidize all unsheltered industries would involve an unbearable drain on the Exchequer; to protect them would raise the cost of living, sheltered wages and the general price level—a change which would handicap still more our export trades. The right and natural way to correct the discrepancy is to lower costs in the sheltered industries, by increasing output where that is possible, and, where it is not, by lowering nominal wages into conformity with the expected new value of money, thus keeping real wages approximately constant. It would be still better to have a stabilized currency and avoid these industrial dislocations altogether.

The pound is now fixed in terms of gold. We may expect the demand for and the supply of gold to vary as in the past, and slow changes in the price level to follow. But we need not fear a recurrence of the sudden and great monetary alterations of the last few years. If costs can once be brought down proportionately to the prices of agricultural produce, they need not get so badly out of proportion again. But to bring them down involves an increase in output or a lowering in nominal wages in sheltered industries, and that needs an understanding with the trade unions.

And now a few words must be said about the short-term fluctuations of good and bad trade, recurring at intervals of eight to ten years, and affecting agriculture as well as other industries.

Much research has been made by economists of late into the causes of this trade cycle, and into the possibility of its control. It seems to be due to a combination of economic, monetary and psychological factors.

Perhaps some discovery or invention or a mere change in economic conditions starts a new industry, or cheapens and expands an old one. The wealth thus brought to some men may quicken the circulation of currency, and raise prices. In turn, higher prices call for more currency and credit. An idea of coming prosperity gets abroad. People hasten to buy, and embark on new ventures. The banks expand their loans and thus create new credit and swell their deposits. Prices rise further; the process becomes cumulative, and we are swept upwards on what is called a boom.

Then something happens. Perhaps production runs ahead of either demand or currency, which controls purchasing power; stocks accumulate and prices have to be lowered to clear them. Perhaps the banks get alarmed, and raise rates of interest and become cautious about fresh loans. Possibly there is a drain of gold to some foreign country from London, and Bank Rate is raised to check the movement. Everyone expects a fall in price, so no one will buy and thus the fall is hastened and increased. The boom is followed by a slump.

Some economists hold that the extreme fluctuations thus created may be diminished in intensity now that their causes are understood. One factor in the cycle at all events is under control—Bank Rate. In pre-war days, Bank Rate was adjusted chiefly or entirely in view of the gold reserve of the Bank of England. If gold was exported too much, the rate of discount was raised, and thus foreign remittances attracted to London. When gold arrived, the rate was lowered again. By this means our foreign trade was kept in a healthy state. When our exports, visible and invisible, were less than our imports, gold was sent abroad to pay for the excess. Thus an export of gold was a sensitive sign that our prices were too high.

The process of adjustment in old conditions is well seen by turning to a description, given in the *Report* of the Cunliffe Commission in 1918, of the way that prices were brought down before the war by the Bank of England.

The raising of the Bank's discount rate and the steps taken to make it effective in the market necessarily led to a general rise of interest rates and restriction of credit. New enterprises were therefore postponed and the demand for constructional materials and other capital goods was lessened. The consequent slackening of employment also diminished the demand for consumable goods, while holders of stocks of commodities carried largely with borrowed money, being confronted with an increase of interest charges...and with the prospect of falling prices, tended to press their goods on a weak market. The result was a decline in general prices in the home market which by checking imports and stimulating exports corrected the adverse trade balance which was the primary cause of the difficulty.

Thus the process worked by lowering the internal price level at the cost of "slackening of employment," and the financial embarrassment of those who had "to press their goods on a weak market." If human interests and human nature were not involved in the details of the adjustments, the process would remain a beautiful example of economic relations and an effective method of control. But its beauty and effectiveness depend on trade losses and on the lowering of money wages to meet slackness of employment, and, though falling prices may supply the appropriate set-off, and prevent an equal fall in real incomes, every growth in the power of trade unions makes the necessary lowering in nominal wages more difficult.

But, even in present conditions, some effect can be produced. The discount rate of the Bank of England usually determines the interest charged by the Joint-Stock Banks for loans to their customers, and hence the facility of borrowing for the purposes of business. Its power is limited, but still doubtless it does control to some extent the amount of business undertaken. Hence, by raising Bank Rate early in a boom, something might be done to check it, and, by making loans cheap in a slump, its evils might be lessened and curtailed. Whether this action can be made effective with due regard to the stability of our banks, whether indeed it can be combined with the action necessary to maintain a gold standard, now that we have recovered that mixed blessing, is still a subject of discussion amongst bankers and economists. It is worthy of note that the International Conference held at Genoa in April 1922 recommended the summoning of a meeting of representatives of central banks for the purpose of arranging common action designed to check fluctuations in the purchasing power of gold. This policy was endorsed by all the governments concerned, but no steps have been taken to carry out the agreement.

The result of this analysis is to show that, while individual changes in the price of crops may depend on the chances of plenty or scarcity in badly organized markets, the fluctuations of the trade cycle are due to a combination of factors, not primarily concerned with agriculture. Again, the broad and long-continued changes which bring general agricultural prosperity or adversity are due not to agricultural but to monetary causes. The growth of business, sometimes hastened by expanding population, sometimes by new discoveries or inventions, needs a corresponding growth in currency and credit. If this lags behind, prices fall, and, while the fall continues, farmers cannot make profits, and agriculture must be depressed.

The analogy of the tide, with which we began our analysis, is seen to be a true one. Watching by the sea the individual waves, we fail to note its slow but irresistible ebb and flow. And, concerned directly with the buffeting of individual dealers and markets, the farmer does not realize the slow change in the value of money, which controls inevitably the price of world commodities. Other men, perhaps equally afflicted, seem able to sell at figures to him unremunerative, and so the underlying monetary cause appears in the guise of unfair foreign competition. Unfair indeed it is if his currency is rising in value compared with those of foreign countries, or has risen without a corresponding fall in the prices obtained by sheltered industries, the goods or services of which enter into his costs of production. On the other hand, if, by a scarcity of gold, the currency of nearly all countries is appreciating together, foreign farmers may equally with British be the victims of falling markets, and each may blame the other for over-production and cutting of rates, when the real cause is the irresistible ebb of the general price level owing to an insufficient supply of gold and of the currency and credit based on it.

Just lately foreign farmers have had an advantage,

because our currency has been rising in value compared with the American dollar or the Danish crown. But foreign competition is not the primary cause of our farm losses. It is an effect produced by the same cause. The common cause of these losses and of foreign competition, as in the years 1874 to 1896, is to be sought in monetary changes, partly inevitable, and partly the effect of a deliberate financial policy for which all political parties in this country are almost equally responsible.

Deflation and a fall in prices from 1920 to 1923 was necessary if the nation was to regain control of the currency. The further deflation of 1924 and 1925 and the return to the gold standard at the pre-war value of the pound may possibly have been an advantage to the nation on balance—it is difficult to judge. But nothing is gained by shutting our eyes to the certain fact that it inflicted grave injury on the unsheltered trades, and not least on agriculture.

A fall in price due to the real cheapening of an industrial process by improvement or invention is a great gain to the world. But a fall due to monetary contraction, which reduces purchasing power, tends to make industry unremunerative, and leads to depression and unemployment.

The discovery in the near future of a large new goldfield followed by a natural expansion of currency and credit is unlikely. The best chances of agricultural recovery depend on the international stabilization of the general price level, combined with a growing shortage in the world output of food at present prices.

# PART II

## THE LAND AND ITS OWNERS

### CHAPTER VI

### THE HISTORY OF THE MANOR

No proper understanding of the present conditions and future possibilities of country life can be won without an appreciation of the past history of English land. Of recent political pronouncements, that of the Liberals takes into account the development of the modern agricultural estate from the mediaeval manor, but it can best be studied in other books, such as Lord Ernle's *English Farming Past and Present*[1]. It is unwise to simplify the story too much, but, nevertheless, since nowadays there is much talk about the iniquity of enclosure, it will be well to recognize the economic causes which underlay the long process leading up to the Enclosure Acts, and learn not to fall into the common fault of indiscriminate denunciation.

Most feudal landlords held on condition of military service; the money payment of soccage, or special services other than military, were far less common. The Liberal Land Book lays stress on the doctrine of the Common Law that land, even when in fee simple, is not in absolute ownership, but is still nominally held of some overlord, and ultimately of the Crown. It says that the principle of national service for the use of land was never

[1] First edition, 1912; third edition, 1922.

abandoned. It omits to mention that feudal services were abrogated by the Commonwealth, and deliberately not reimposed at the Restoration. Still, I have no quarrel with the view that landownership is a form of national service, though I doubt the assumption that the landowner is now ceasing to render it. Perhaps it is allowable to point out that he does more work for his rents than the holder of stocks and shares does for his dividends. I think the idea of a call to national service might well be extended to all forms of property.

As the lord held of an overlord or the Crown, so the tenants of the manor held of the lord by various services—usually a stated number of days' work on the lord's demesne—which were later commuted for money payments, the origin of copyhold rents. The tenure of the holdings, the fines for renewal, etc., were settled by the custom of each manor, custom which, as interpreted by the Manorial Courts, acquired the force of law.

In these courts the life of the manor was governed. The lord or his steward presided, but freeholders and copyholders took part in the decisions. Succession to holdings was regulated and recorded; offenders against property or person or custom presented, and fined or otherwise punished, and a certain standard of cultivation maintained.

Land was either enclosed or in open field. The winding, often deep-set, country lanes, seen in parts of the West, East and North of England, still show us the lines of early British tracks, and the irregular little fields show the closes taken in by individuals directly from the waste. Open fields, cultivated in common by the tenants and lord of the manor, by the eighteenth

century were only found in a wedge-shaped tract of
country covering the South, the Midlands and part
of the North, though many who use the history of
enclosure for political purposes assume that they were
universal. In open-field villages, the rigid framework of
the rest of mediaeval life extended to husbandry also.
Each tenant held strips of land scattered in (usually)
three immense fields, and, from year to year, had to
treat each strip as the other tenants did the rest of the
field. Such a system, while it brought pressure to bear
on a tenant who openly lagged behind the average
standard, gave no scope for individual enterprise or
initiative, and led to a deplorably low level of cultivation
and output of food.

Hence enclosure both of open arable fields and
commons or wastes was always going on, usually slowly,
but sometimes rapidly, for special causes like the development of the wool trade and other economic changes
from 1485 to 1560.

At this time some land was enclosed to convert open
arable field into sheep pastures, which needed less
labour, and enclosure got a bad name in consequence.
The process met with fierce denunciation from the
pulpits of Romanist and Reformer alike, and the determined opposition of Tudor governments concerned
to prevent the depopulation of the countryside.

Though already breaking down in practice, the
mediaeval scheme of life still held men's minds. Society
was regarded as a rigid framework of law and custom,
part of the complete religious synthesis of the School-
men. Within this fixed framework the economic motive
had to act. The scheme was meant to give protection to

each man in his allotted station; it did not contemplate changes by which, through individual advancement, wealth might increase, and the general standard of life be raised.

In England this older social theory passed away amid the flux of the Civil War and the Commonwealth. With the rise of a commercial class to power, it had come to be assumed that the economic motive moulded the framework of society, and that, except for the legal prohibition of definite, prescribed crimes, each man's private conscience had to adapt his own actions to the requirements of "business." The intellectual position had been completely reversed. Economic forces, however softened in practice by kindly human relations, in theory worked almost unchecked by official moral bounds from the breakdown of the personal government of Charles I, Stafford and Laud, till the days of Lord Shaftesbury's Factory Acts. The industrial revolution, though its evils compared with those of the preceding age have sometimes been exaggerated, worked its will unbalanced and uncontrolled, leaving a legacy of misery and misunderstanding from which we still suffer[1].

Enclosure went on, though more slowly. But now corn was needed, and with the introduction of turnips, sheep could be kept on arable land. Enclosure no longer meant direct depopulation. The wasteful open fields became fenced plough lands, and grew far better crops. Therefore, when the growing population and the demands of war called urgently for more food at the

[1] For an account of the change from mediaeval to modern theories of life, see R. H. Tawney, *Religion and the Rise of Capitalism*, Murray, London, 1926.

end of the eighteenth and beginning of the nine-teenth centuries, local and then general Enclosure Acts allowed the change to be carried out rapidly. The common arable fields were apportioned among the tenants of the manor, and some rights of common on the waste extinguished for money payments.

Individuals thus obtained compensation, and no injustice was intended, though some of the consequential effects were socially disastrous. It is now usual to represent enclosure as a robbery of "the people" by the landlords. This view does not survive a study of the facts and figures. As Lord Ernle says[1]:

The object of enclosure was not to effect any transfer of ownership. The men who proved themselves to be owners emerged from the process as owners. In its immediate effect, enclosure rather tended to increase than diminish the number of freeholders, for it recognized, in certain cases, the claim of copyholders, leaseholders and squatters to a freehold interest in land. What it did was to change the subject-matter of the property owned, to substitute a compact block of freehold land for common rights, and to make the change compulsory.

That enclosure did not of itself create large estates is shown by the Return of the Enclosure Commissioners in 1876[2]:

Between the years 1845–75, 590,000 acres were enclosed. They were divided among 25,930 persons...620 lords of manors received, on an average, $44\frac{1}{2}$ acres each; 21,810 common right owners received, on an average, 24 acres each; 3500 purchasers (of land sold to pay the expenses of enclo-sure) received, on an average, 10 acres each. Among the 21,810 common right owners were 6624 shopkeepers and tradesmen, labourers and miners.

[1] *The Land and its People*, p. 57.
[2] Quoted in *The Land and its People*, p. 59.

Of course serious errors may lie hidden in an average. But the division of 590,000 acres of land among 25,930 persons between 1845 and 1875 is a monumental fact. The proportions are not likely to have been very different in the earlier years from 1760 onwards. It was not enclosure, but its indirect consequences in the changing economic conditions of the day, that led to consolidation. Land, freed of mediaeval restrictions, became easier to deal with and worth more to sell. The sum paid for the extinction of rights of pasturage on the common waste may have been adequate compensation to the individual, but it turned him and his heirs from their accustomed mode of small husbandry.

Simultaneously, village industries were vanishing under the competition of cheap factory-made goods, and small men, who had partly depended on them for a living, faced with the immediate expense of fencing their allotment, often sold their land. Doubtless they formed a goodly part of the 3500 who had transferred their allotments to the "purchasers" enumerated in the Return of 1876. As time went on, more of the "21,810 common right owners" followed their example, and the number of separate ownerships was diminished as an indirect result of the long process of enclosure.

With the purchase price of their land, many copyholders and yeomen embarked on manufacture in the towns growing so fast at the end of the eighteenth and beginning of the nineteenth century. Some of them became leaders of industry, founding great commercial families; others, realizing their fortunes, returned to the country as landowners on a larger scale. Yet others failed to make good, and swelled the ranks of rural or urban labourers.

Though the number of holdings in England is still more than usually supposed[1], the final effect of enclosure on the countryside was to decrease the number of copyholders, other customary tenants and small freeholders. The standard of farming and the output of food were raised enormously—indeed without enclosure the country could not have been fed during the Napoleonic Wars, and modern high farming would have been impossible. But the mediaeval system of country life was finally broken down, and the present system of landowner, tenant farmer and labourer emerged as the predominant feature of social life over the rural part of southern England, though numbers of smallholders and occupying owners survived all changes.

To this process of evolution land reformers are now accustomed to assign the blame for what they call "landlordism," as well as for the existence of the class of "landless labourers" and for the low wages they obtain. Doubtless it is true that in some other countries more smallholdings have survived the changes which began when the land was enclosed. But the class of "free" (*i.e.* landless) labourers appeared in England in early mediaeval times, and there was a less sudden change than the rediscoverers of the evils of enclosure believed. Change was going on throughout the Middle Ages; it is difficult to judge fairly the economic welfare of the serf or the villein or the copyholder at different periods, and those who know most seem nowadays least

[1] In 1921 there were 420,133 holdings in England and Wales with an "average" area of 62 acres. See p. 195 below.

inclined to dogmatize. Mr G. M. Trevelyan, in a recent book says[1]:

> No doubt the period of the two first Georges, with its good wages and moderate prices, compared favourably with the period of rural pauperism in the early nineteenth century. But there had been hard times before, in days when hard times meant famine. In the "dear years" of William III, and often before, people had failed to "subsist" on their "subsistence agriculture." The cottars, too, whose disappearance we deplore, had been classed with the paupers by Gregory King, a publicist of William's reign. King's often-quoted analysis of English society at the time of the Revolution, points to the existence of a rural proletariat more numerous than the yeomen and tenant farmers put together.

We are sometimes apt to exaggerate the advantages of old times because they have passed away. In spite of all that is said, one thing is certain—whatever was the peasant's position in the mediaeval manor, and under William III or the early Georges, the labourer's life is far better now than it was a hundred years ago and down through the "hungry forties." *E pur si muove.*

Nevertheless, no one can study the Court Rolls which concern a parish he knows without seeing that the slow decay of the manor did unwittingly destroy the old social life of the village. The break was more sudden in open-field parishes, where enclosure, necessary as it was, produced evils unintended and unforeseen. But, even in early enclosed counties, there was a slow consolidation of holdings, involving perhaps no hardship to individuals, but still resulting in a decrease in the number of copyholds and small freeholds. And everywhere village industries,

---

[1] *British History in the Nineteenth Century,* p. 8.

on which many cottars partly lived, were destroyed by the competition of the new factories and improving transport which brought their products to the door. The other functions of the Manorial Courts have been gradually transferred to the more centralized jurisdiction of Assizes, Police Courts, Inland Revenue Offices and County Councils. This was inevitable. The manor was the natural unit of administration in mediaeval times, but it is far too small with modern industry, with modern communications and indeed with modern life generally. The manor as a self-governing community slowly passed away, and much of the natural social life of the village went with it.

By means of Smallholdings, Football Clubs, Women's Institutes and Rural Community Councils, we are trying painfully to put something together again; but, with the real business and local administration of the village perforce excluded, no complete remedy is possible. No shadowy control by a County Agricultural Authority, such as is suggested by Liberal and Labour reformers, will supply the local knowledge and personal touch of the homager in a Manorial Court dealing with the affairs of his own parish, especially with the joint cultivation of common fields, in which his neighbour's strips lay unenclosed next his own. To elect representatives on a County Committee is a poor substitute, and to suggest it as a cure for present deficiencies is to misjudge entirely both the lessons of the past and the possibilities of the future.

## Chapter VII

## THE LANDOWNER

Though the change was less sudden than is often made out, and many causes combined in the evolution of our present rural society, there is no doubt that the enclosure of the common arable fields helped the larger landlords in parts of England in the eighteenth and early nineteenth centuries to consolidate their estates and obtain a predominant position in the countryside. It was not the only cause. The growing wealth of the nation enabled energetic landowners to increase their fortunes in various ways, and gave good chances for rising men from other classes to expand industry, make money, and buy land.

Many hard words have been said about the changes thus brought about, sometimes by the very men who blame present landowners for losing that leadership held by their great-grandfathers. For instance, the Liberal Green Book in one place traces all our troubles to this establishment of "landlordism," and, careless of consistency, in another sings the praises of the very embodiment of its spirit in the "great improving landlords" of the time of the enclosures[1].

Landlordism in the nineteenth century owed much also to the lingering memories of the great reforming agriculturalists of the previous century. "Turnip" Townshend, the Duke of Bedford, Lord Egremont and pre-eminently Coke of Norfolk, who died only in 1842, landowners all, had displayed a real genius of enterprise....Devotion to the land,

[1] *The Land and the Nation*, p. 226.

real knowledge and understanding of it, the lack of competing interests, and ample resources, were the qualities which enabled Coke and those who rank with him to succeed.

The same point had previously been put by W. E. Lecky when he said:

It is impossible to consider the history of English agriculture in the last century without arriving at the conclusion that its peculiar excellence and type sprang from the fact that the ownership and control of land were chiefly in the hands of a wealthy and not of a needy class.

It is argued that the system of land tenure in this country has broken down mainly for two reasons: firstly, because landowners have ceased to lead the farming community as scientific agriculturalists; and, secondly, because they are now too poor to maintain buildings, drainage, and the other permanent capital equipment of the land.

Let us deal with these two criticisms of the landowner in order.

1. *Leadership*. In discussing the depression of 1874 to 1896, the Green Book says[1]:

In Great Britain the need for a national movement in support of agriculture was not recognized. The country as a whole, having a generation before lifted its eyes from the furrow to the factory, assumed, so far as it bothered about agriculture at all, that the system which had served the needs of agriculture in good times might serve it in bad. This was a fatal miscalculation. What was required was a fresh impetus, some force comparable to that exerted at an earlier date by Townshend, Egremont, Coke and Sinclair. It was not given by their successors,

either then or at the present time. Here again it will be seen the authors of the Green Book, like other critics,

[1] Pp. 243, 244.

ignore the true causes of agricultural depression. The high farming of the "improving landlords" would not have prevented bad times. Nevertheless, their race is not extinct, for on the very next page the book supplies an answer to its own complaint:

Happily there are now in England a certain number of landlords who are themselves admirable farmers. Lord Bledisloe, Mr Ismay, Lord Guilford, Lord Folkestone are the first names to come to mind.

It is amusing to observe that even "the first names to come to mind" among the landowners who are "admirable farmers" to-day, are equal in number to all the "improving landlords" that the authors could remember during the whole hundred years of the "Golden Age." That indication alone suggests that things are not so bad as the Green Book likes to make out. It states that leadership has now passed from the landowner to the agricultural expert and the County Agricultural Organizer. In regard to the more technical scientific side that is true—but also inevitable. As in the other sciences, advance in agricultural research has almost ceased to be possible to the amateur or even to the professional farmer. It needs the resources of a laboratory with costly equipment and extensive staff, worked in conjunction with an experimental farm with ample financial resources. Similarly, the amateur physicists, chemists and biologists of fifty or a hundred years ago, are now replaced by highly trained specialists, working in elaborate university or national laboratories.

To expect all or most landowners to do this work betrays confusion of thought. Such highly technical research must in modern times be organized carefully on the large scale. But the book is really confusing it with the different work of applying its results in practice, raising the standard of progressive farming, making the new knowledge available to ordinary farmers, and inducing in them a receptive attitude of mind. In this work, I venture to think, more landowners are taking part, or preparing themselves to take part, then ever before.

I have some knowledge of agricultural societies and of schools of scientific agriculture; yet if I pit my mere impressions against those of the investigators who supplied the data for the Liberal Land Committee, I shall have no better right to speak. Vague impressions are poor things on which to found a policy. I have therefore obtained facts and figures from two large Schools of Agriculture—those of Oxford and Cambridge. Professor Watson of Oxford has kindly analysed the list of his 67 students in 1925–6, and finds that 33 are the sons of landowners, and of these 13 or 14 are known to be likely to succeed to estates of considerable size in Great Britain. At Cambridge, from information supplied by the Tutors of all the Colleges, it appears that, out of 162 students of agriculture, 42 are the sons of landowners, and 35, more than one-fifth of the whole, are owners or heirs of considerable estates. Thus at the two Universities, about 50 men are training themselves seriously for the ownership of land, besides an uncertain number more, let us guess about 20, at the various Agricultural Colleges—perhaps about 70 in all. Of each generation of 33 years, the three years of a University Course form one-eleventh part, so it appears that, at the present rate, the heirs of about 700 or 800 estates will have taken an agricultural course before they succeed their fathers. There are about 2000 peers and baronets, and Burke's *Landed Gentry* contains entries relating to about 2000 additional names. We may perhaps estimate the whole number of families in possession of considerable estates as somewhere about 3000 or 4000. Thus the estates with future owners who have been through a scientific course in agriculture constitute a considerable fraction of the whole number. Indeed, the proportion is surprisingly high when it is remembered that many eldest sons now take up some other profession during at all events their earlier years. In the interesting historical sketch he gave to the British Association in 1926, Professor T. B. Wood emphasized the importance of the discovery by the sons of landowners that agriculture could be learnt at Cambridge. This discovery was one of the chief factors which gave an impetus to the original small-scale school, and enabled those concerned, with the support of the Ministry, to develop it into the great department which has done so much for scientific agriculture.

2. *Poverty.* The contention that landowners are now poor, possesses of course much truth, though in most cases it is their personal expenditure and not the equipment of their estates that has suffered. Death duties, which have been levied already two or three times on most estates, make it difficult for a family to remain long in possession of their ancestral acres, unless they economize stringently or somehow bring in new wealth. The cost of the upkeep of buildings, and on heavy land of drainage, has reduced the net returns from agricultural estates to a small fraction of the gross rent roll,

a fraction that is itself heavily depleted for income tax. Unless the land is exceptionally fertile, often little remains for the owner. The Green Book gives examples of estate accounts which show that net returns before the war varying from 52 to 25 per cent. of the gross rentals, had fallen in 1924 to percentages varying from 30 to 16. These figures are before deduction of income tax. When that is allowed for at present rates, it is not surprising that the book can give an instance of a large property where "the landlord's personal income from the estate...was about enough to pay the tax on a medium power motor-car." Evidence given below shows that these net returns are lower than the average, but they illustrate well the harder cases.

Into the question of death duties I do not propose to enter at length. Where properties consist of stocks and shares, death duties may work fairly well, but to land-owning families they are the worst and most unfair of all forms of taxation. To pay income tax is unpleasant; but, after all, the business of the country must be carried on, and it is right that each man should contribute in accordance with his personal income—to do so should leave no sense of grievance. But death duties confiscate a part of property which morally belongs to the family and not to the individual; they confiscate it each time the misfortune of death occurs, and on a scale which, based on an estimate of what the estate would fetch if sold farm by farm with no allowance for the expenses of sale, is much higher than the capital value of the real income obtained. Such an exaction, falling at incalculable intervals, is grossly unfair as between one family and another whose changes and chances of mortal life happen to be fewer. It also impoverishes periodically the estate and all that dwell thereon, and is directly destructive of that stability on which healthy country life depends. As the Green Book says[1], "It is difficult to make a success of the British system of tenant-farming unless the landowner can afford to take a long view and follow a continuous policy." Many improvements will benefit chiefly a man's son or his grandson, and it is the shadow of death duties more than anything else which makes it impossible to take the long view the book rightly praises. Each time the duty is exacted it leaves behind a bitter sense of injustice, the mark of a bad tax. If it be necessary to maintain the duty in general, I believe that it would be wise as well as fair to charge estate duty on agricultural land on the capitalized value of the net income therefrom until the land is sold, when the balance might become payable. This arrangement would give no advantage to land; it would only place it on a level with other property. The estate would pay duty on the actual and realized value, whether for holding and letting, or for sale.

Death duties are not the only cause of the financial troubles of the landowner. The monstrous charges now exacted for building, and the high cost of other estate repairs, the rise in taxes and rates, and sometimes the burden of a house and garden too large for modern conditions, contribute to his embarrassment. While his estate outgoings have increased by amounts varying from 70 to 100 per cent., his nominal gross rents are either unchanged or have risen by perhaps 20 per cent., leaving a very small net return even measured in

[1] Page 227.

pounds. If, moreover, we take the index number of agricultural and other wholesale prices to be now about 150, compared with a pre-war 100, it is clear by how

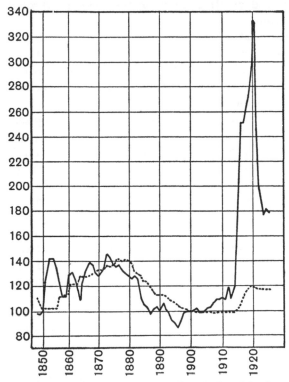

Diagram III. Average rent of agricultural land.
Prices of agricultural produce —— Agricultural rents - - - -

much nominal rents should have increased to have kept pace with the increase in the tenant's gross receipts, or with the fall in the value of money, which has now only two-thirds of its former purchasing power. Allowing for this change, we see that, on their present rentals, land-owners are taking much less real rent than before the war, and are thus carrying the industry through its bad time.

The relation between farmers' gross receipts and the average rents they pay is shown by Diagram III, taken from the *Report* on the Stabilization of Agricultural Prices[1], the curves being extended to cover the last two years on the same basis.

The continuous line gives the prices of agricultural produce when those of 1900 are taken as 100, and the dotted line the corresponding variation in rents, based before 1914 on the income tax assessments. From 1875 to 1900 rents lagged behind the fall in prices[2], and from 1900 to 1920 behind the rise. It will be seen how small was the rise in rent from 1914 onwards compared with the rise in prices by which the tenant benefited.

But, if we are to compare fairly the effect of price changes on the three partners in the agricultural industry, these crude figures need analysis and correction. A farmer's net profits do not rise in proportion to his gross receipts. Diagram II on p. 66 gives the average profit on an ideal arable sheep and corn farm, and a similar diagram has been drawn for a grass dairy farm[3]. The mean value of the profits on these two different farms will give a

[1] *Report of the Committee on the Stabilization of Agricultural Prices*, p. 12.

[2] Basing the curves on income tax assessments does not allow for temporary remissions of rent, which are customary in the early years of a depression. Permanent reductions of rent come later. Thus the curve exaggerates the lag in times of falling prices.

[3] *Journal of the Royal Agricultural Society*, vol. LXXXV, 1924, p. 137.

not unfair representation of the average financial results
of farming in the South of England during the years
1914 to 1925.

Similarly from the landowner's rents must be de-
ducted his outgoings to find his net receipts. The estimation of the increase in rents quoted above seems based
on inadequate data. A new enquiry has therefore been
undertaken, the full results of which will be set forth
elsewhere when the work is done. The figures at present
obtained are those from three small private estates,

TABLE III

*Rents and Outgoings of Agricultural Estates* 1913–1926

| | Gross rents. 1912–1914 = 100 | | | Outgoings as percentages of gross rents | |
|---|---|---|---|---|---|
| (1) Year | (2) Three small private estates | (3) Three large corporate estates | (4) Mean of (2) and (3) | (5) Of each year | (6) Of 1912 –1914 |
| 1913 | 100 | 100 | 100 | 32 | 33 |
| 1914 | 100 | 100 | 100 | 33 | 33 |
| 1915 | 100 | 101 | 100 | 34 | 34 |
| 1916 | 100 | 101 | 101 | 35 | 35 |
| 1917 | 100 | 102 | 101 | 36 | 36 |
| 1918 | 105 | 103 | 104 | 37 | 38 |
| 1919 | 107 | 105 | 106 | 44 | 46 |
| 1920 | 109 | 108 | 109 | 57 | 61 |
| 1921 | 126 | 116 | 121 | 61 | 74 |
| 1922 | 129 | 120 | 125 | 58 | 72 |
| 1923 | 123 | 119 | 121 | 47 | 57 |
| 1924 | 122 | 120 | 121 | 47 | 57 |
| 1925 | 122 | 120 | 121 | 47 | 57 |
| 1926 | 120* | 120 | 120 | * | * |

* Returns not yet complete.

about 6300 acres in all, and three large estates belonging
to corporate bodies, comprising some 35,000 acres. The
land is well scattered about England. It will be seen
from Table III that there was very little change in rent
during the war. From 1920 to 1922, it seems that the
small private owner received proportionally rather
more than the large corporate body, but quickly reduced
his rents again when prices fell. Taking the mean between
the figures for these two groups of properties, we get
Column 4 in the table. The outgoings are more difficult
to calculate, and at present only those of the three
private estates and one corporate estate are available.
These are shown in Column 5 as percentages of the
corresponding gross rents, and in Column 6 as percentages of the rental of the years 1912–14.

As approximate values we may take it that the
average money rents of agricultural land rose by
25 per cent. between 1913 and 1921 and have now fallen
to an index of about 20 per cent. Outgoings before the
war were about 33 per cent. of the 1913 gross rentals.
This figure rose to about 74, and is now about 57. On
the rentals of each corresponding year the percentages
of outgoings are 61 and 47 respectively.

These are all round numbers, subject to revision, but
they are not likely to be changed materially by further
data. They are probably fair average figures, and, although they show a considerable reduction in net
income (see Table IV), it will be seen that they are not
nearly so bad as examples which have often been given
from the more unfortunate estates both by landowners
and their critics when, for different reasons, they have
wished to prove poverty.

Having dealt with profits and rents, we must turn to wages. The average wages of farm labourers from 1917 onwards have been taken from the annual statistics compiled by the Ministry. From 1914 to 1917 they have been assumed to change proportionally.

By these methods we get figures for the alterations in money profits, rents and wages during recent years. But the important thing is not the number of pounds we have; it is the amount of goods and services those pounds will buy that constitutes our real income.

Farm labourers receive some payment in kind. A Committee on the Occupation of Agricultural Land[1] estimated the labourer's cash expenditure at two dates in 1918 as 86 and 93 per cent. above that in 1914, an average increase about 16 per cent. less than that given by the Board of Trade cost of living for wage earners generally, which showed increases of 95 and 120 at the same dates.

The necessary expenditure of landowners and farmers has not been studied, but Professor Bowley estimated that of doctors in 1923, when special war conditions had ceased, as 57 per cent. above 1914, when the Board of Trade cost of living was 69. This difference of 12 is 17 per cent. of 69, the same as the 16 per cent. for the labourer within the somewhat wide limits of error.

We shall probably not be far wrong if we take the increase in the cost of living for all three partners in agriculture at 16 per cent. below that given by the Board of Trade index, though it must not be forgotten that the calculation takes no account of income tax, which depletes the net income of landowner and farmer

[1] Cmd. 76. 1919.

much more now than it did before the war. Table IV
sets forth the results, and Diagram IV shows them
graphically.

TABLE IV

*Purchasing Power of Agricultural Profits,*
*Wages, and Rents*

| Year | Cost of living for agricul- turalists | Profits | | Wages | | Net receipts from rents | |
|------|------|-------|------|-------|------|-------|------|
| | | Crude | Real | Crude | Real | Crude | Real |
| 1913 | 100 | 100 | 100 | 100 | 100 | 100 | 100 |
| 1914 | 100 | 135 | 135 | 100 | 100 | 100 | 100 |
| 1915 | 119 | 240 | 201 | 117 | 98 | 99 | 84 |
| 1916 | 139 | 380 | 273 | 134 | 96 | 99 | 71 |
| 1917 | 164 | 535 | 326 | 150 | 91 | 97 | 60 |
| 1918 | 187 | 555 | 297 | 194 | 103 | 98 | 52 |
| 1919 | 197 | 590 | 299 | 236 | 119 | 90 | 45 |
| 1920 | 225 | 630 | 280 | 281 | 125 | 72 | 32 |
| 1921 | 206 | 125 | 61 | 263 | 128 | 70 | 34 |
| 1922 | 170 | − 50 | − 29 | 200 | 118 | 79 | 46 |
| 1923 | 162 | 120 | 74 | 175 | 108 | 96 | 59 |
| 1924 | 163 | 215 | 132 | 156 | 96 | 96 | 59 |
| 1925 | 164 | 185 | 113 | 174 | 106 | 96 | 58 |

The table and the diagram show that the farmer
increased his purchasing power greatly from 1914 to
1920, but suffered loss in the post-war slump. The
labourer found his real wages falling slightly in value
till 1917. They then recovered, till in 1921 they were
higher than in 1913 by 28 per cent. They then fell again
till 1924, as money wages were reduced faster than the
fall in prices. Since the re-establishment of Wages
Committees, real as well as nominal wages have risen,
and in 1925 were about 6 per cent. above pre-war

values. It must not be forgotten that hours of work are shorter than before the war, and the value of national services such as old age pensions, children's education, etc. has increased. On the whole, the labourer's position has improved.

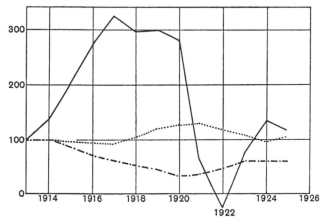

Diagram IV. Purchasing Power of Agricultural
Profits —     Wages ......     Net Rents — · —

Rents fell in purchasing power from the beginning of the war till in 1920 on the average they would buy only about one-third of what they would buy in 1913. Many estates brought in no return, or even involved a loss to their owners. Thereafter real rents recovered as prices fell, and they seem now on the average to be about 60 per cent. of their pre-war values. The table and diagram of course take no account of income tax, which reduces further both the farmer's profits and the landowner's rents.

w                                                                 8

Thus the present position seems to be that profits and wages possess a little more than their pre-war purchasing power, wages tending upward and profits, for the time at any rate, downward. Rents possess about 60 per cent. of their pre-war real value, with some signs of a renewed fall.

The landowner has been badly hit. He is recovering, and is not in quite such a parlous state as some kind friends who await his decease make out. But he is still suffering from the shock of war.

The meaning of these results is clear. Besides the open contribution to the cost of the war levied by income tax, an additional covert agricultural contribution to the cost of the war and to the exactions of the building trade levied on the nation during the peace is being taken from the industry by reducing the landowner's real income. At the owner's expense, the average farmer's profits are being subsidized, and are now oscillating about pre-war values, while the labourer has slightly gained in the purchasing power of his cash wages, and in other ways has considerably improved his economic position. This is, happily, in marked contrast to the years which followed the Napoleonic Wars, when some of the loss fell on the labourer and his life became deplorable.

Thus we return to the statement already made above. By voluntarily or involuntarily taking lower real rents, landowners are now bearing the chief burden of the agricultural depression, and carrying the industry through its troubles. To act thus as a buffer to absorb economic shocks is one of the functions of the landowner. The

fact that he is performing this function explains why
he is now so poor. But it also explains why agriculture,
with all its troubles, is in far less parlous case than are
some other "unsheltered" industries. The indications
of our diagram are confirmed by Mr Venn's investigations into the actual accounts of a number of farms.
Before the last rise in the value of the pound and in the
rates of minimum wages renewed the depression, even
in East Anglia, with these low rents, farming was again
beginning to pay[1]. The farmer's temporary balance,
however, seems now likely to be absorbed, partly by a
rise in agricultural wages, which no one grudges, and
partly by the subsidy which the return to the gold
standard gave to the sheltered at the expense of the
unsheltered industries. This has gone to raise real profits
and real wages in trades that did not need help, and has
depressed further those that were already in trouble.
The average index of agricultural prices for the first
three months of 1926 was 53, as compared with 69
for the corresponding months of 1925 and 59 for the
whole of that year.

When, however, the process of readjustment is
finished, either by a rise in world prices or a reduction
of wages and other costs in our sheltered industries,
such as building, transport, etc., real rents should again
bear something like their old ratio to outgoings. The
landowner will then once more be able to maintain the
fixed capital of agriculture without the present ruinous
strain on his resources.

Of course the nation has a right to buy out the owners

[1] *Farm Economics Branch, Cambridge School of Agriculture,*
*Reports 1 and 2.*

on equitable terms if Parliament decides to take over any kind of property, but it is indeed hard that this moment should be chosen to launch a scheme for expropriating the landowner. To buy him out compulsorily just now, when, willingly or unwillingly, he is accepting an abnormally low rent, and to pay him compensation on that basis, as is proposed, is clearly unfair.

One land reformer, Mr Philips Price, writing, I imagine, from a "Labour" point of view[1], thinks that even this compensation on the basis of the net rent at the time of expropriation may turn out to be too generous. He calmly proposes that, until some scheme of nationalization can be carried through, the County Committees should control rents, to "prevent landlords from profiting by an improvement due to increased prosperity in agriculture." Verily the country landowner can expect neither comprehension nor justice. He is now carrying the chief burden of agricultural depression in the hope of better days, and even this hope is to be taken from him.

Let me add at once that the scheme put forth in *The Land and the Nation* was acknowledged to be unfair in its terms of expropriation by the National Liberal Federation in February 1926, and a new basis adopted, namely, the assessment of the Valuation Department, presumably the value of the land as assessed for death duties, excluding what was called "monopoly value."

The Conference modified the scheme in another important way. In the Green Book, all agricultural land was to be taken over compulsorily and together at a stated time, and arguments, some of them sound if any

[1] *The Guardian*, January and February, 1926.

such scheme is to be adopted at all, given for refusing
to consider gradual acquisition. All this was thrown
over by the Conference, from whose troubled waters
compulsion and universality emerged as a mere duty
of taking over land needed for specific purposes or
offered voluntarily for sale by its owner.

It must be admitted freely and cordially that these
amendments make the scheme more equitable from the
point of view of the individual landowner and less
dangerous to the nation. Nevertheless, as amended,
there is some doubt as to the meaning of some of the
financial provisions. The Conference deliberately approved the provision that the price paid should exclude
"monopoly value," a phrase that seems to possess
attractions for most land reformers. The inverted com-
mas are copied from the summary of the proceedings
of the Conference, and, I imagine, are the editor's little
way of calling attention to the fact that, at the Conference, not only could no one define what the words
meant, but almost everyone who referred to them agreed
that they were indefinable. One speaker candidly con-
fessed that the real point of the provision was to secure
"that the land should be taken over at something less
than its ordinary open market sale value." This particular speaker thought that, even with the formal exclusion
of "monopoly value," the price could still include sporting, building and amenity values, but I am afraid that
others who supported the clause hoped that it could be
held to exclude all elements of value ruled out in the
Green Book. Whether this be so or not is doubtful, and
it will be well to understand what the book proposed
on this point. The original scheme may be reverted to

even by Liberals under electoral pressure, and a scheme equally dangerous has been sketched in the Labour pamphlet.

By the scheme described in *The Land and the Nation*, the owner is to receive an annuity based, not on the estimated selling value of his land (on which basis, by the way, the nation has exacted one, two or three payments of death duties), but on his present net receipts in a time of depression, and after his rental has been reduced by an amount which will enable his tenants to pay a new minimum wage to be specially fixed with this scheme in view—how fixed is not made clear.

And now let us turn to the scheme of the Labour Party. The Labour pamphlet says:

> The dispossessed landlord would be compensated on the basis of Schedule A, Annual Value of the holding, diminished in the case of the inefficient landlord according as he had failed to maintain the proper condition of the holding.
>
> The most practical way to acquire the freehold would be to give Land Bonds to the owner which would be redeemable by a Sinking Fund provided from the economic rent of the land. In the course of time these Bonds would all be paid off, and the land would be the clear property of the Community.

Here, again, the compensation is based on the net agricultural rent, which determines Schedule A, while death duties have been exacted on the much larger saleable value. But a worse injustice lurks in the second paragraph. If a sinking fund is to be paid from the economic rent, as well as interest on the Land Bonds, it means that the interest must be less than the rents. Though the nominal par value of the Bonds may be the capitalized value of the Schedule A assessment, if the interest is cut down below the existing rate for Government Securities, the real, saleable, capital value of the Bonds is diminished likewise. It means, in effect, that the unfortunate landowner would be paying the sinking fund in order to buy his land for the State from himself.

The difficulties inherent in this question of compensation are well seen when we find that even Mr Orwin and Colonel Peel, with every desire to deal justly, are driven to propose the Schedule A basis by the insuperable complication of valuation. For the same reason, they exclude all land in Urban Districts from their scheme, and refuse to assess building or similar values on land in Rural Districts; such land is to be left for a stated term of years in the hands of its owner for development and then taken over at the new Schedule A value. It will be clear how many artificial lines are here drawn. For instance, there is much agricultural land in what are, for purposes of Local Government, Urban Districts, and much building land in those classed as Rural.

Orwin and Peel propose to pay for agricultural land on the then existing rate of interest—at present 22½ years' purchase of the Schedule A value—with no deductions. Thus the landowner would at least know where he stood. Probably some such basis as this, with a few additional years' purchase added to represent the excess freehold value according to the class of land and the neighbourhood, would be the method of expropriation which was at once the fairest and the most practicable.

There seems to be much confusion of thought about the differences between the selling and the letting value of agricultural land. For example, the Green Book states that agriculture is injured by what it calls "unreal" values, due to monopoly, amenity, prestige, and facilities for breeding and preserving game, though, as it is admitted that present rents are fixed on bare agricultural values, the statement does not carry conviction. It is proposed to "relieve" the land of all these "unreal" values, which explain some of the difference between the agricultural and the selling price of rural land. The owner will be paid for the "real" or agricultural value only. This calm proposal to destroy by a stroke of the pen wealth to the value of untold millions without compensation to its owners, suggests the compulsory purchase of a picture by the National Gallery on the basis of an estimate of the cost of the paint and a fair allowance (doubtless at trade union rates) for the man's time in laying it on.

The difference between saleable value and capitalized net rent varies greatly. Even in a non-residential neighbourhood with no sporting facilities, fertile land or land convenient for markets will sell for a greater number of years' purchase of the net rent than more sterile soil in an inaccessible spot, which may, perhaps, fetch only 16 or 18 years' purchase of its net annual value. This apparent anomaly is due to the fact that, if prices fall and rents with them, the low rent of the bad land will sooner disappear than the high rent of the good land, and, even short of that, will suffer a higher percentage reduction. The greater uncertainty in the rent of the bad land is allowed for by a reduction in its capital value so that it yields a higher rate of interest. All schemes of expropriation based on existing net rents ignore this financial difference in the value of the property acquired, as well as ignoring all "amenity" values.

Some provision is made in most schemes of expropriation for recognizing prospective building value in assessing the landowner's compensation, but one most important point is generally overlooked. Besides the "amenity" group of values, land has a real economic advantage as an investment over a long term of years. It is not, like fixed interest-bearing securities, quite at the mercy of a fall in the value of money. Slowly rents adjust themselves, and, till the disastrous discovery of death duties, landowning families, with economy in bad times, might hope to preserve their lives of usefulness. It is this hope of permanency which for centuries has made men willing to work hard and save to establish their families on the land, and it is this power of adjustment which gives a special value to real property as an investment for Colleges and other long-lived institutions. No Government Bonds are an adequate exchange.

To talk about monopoly is of course ridiculous. There is some element of monopoly when a man wants to build on a particular site, or a railway to run its line through one particular field. But any land reformer can buy as much agricultural land as he wants to-morrow, and buy some of it for less than the cost value of the buildings and other equipment. It still does not seem to be understood that the average rent of agricultural land in England is in reality not rent for land but interest on the cost of equipment. To reclaim the land, fence, drain and provide it with roads, would have meant from £5 an acre, and to put up buildings at least £7

more, even on the costs of fifty years ago[1]; £20 would
be a modest total estimate at present prices. An average
gross rent of (say) 25s. an acre, reduced by outgoings
to something like 15s. to 16s., gives as a maximum a
bare 4 per cent. on the replacement value of the equipment, with nothing for the land itself. It is only exceptionally fertile arable land or old pasture that has much
natural agricultural value of its own, or earns rent as
that term is used by the economist Ricardo, who defined
rent as "that portion of the produce of the earth which
is paid to the landlord for the use of the original and
indestructible powers of the soil." Ordinary agricultural
"rent" is, most or all of it, interest on the cost of
improvement.

It is interesting to study also the freehold value of
English land in this same connection. Twenty-five years
purchase of the net rents is now considered a full price.
At an average net rent of 16s. the freehold value of
agricultural land is about £20 an acre, as near as may
be the replacement value of the buildings and other
equipment, again with nothing for the land itself. Thus
the "monopoly" value proves to be a myth. In buying
average agricultural land a landowner pays for its
equipment, and then lets it at a figure which brings in
a bare 4 per cent. It is not that the price of English land
is too high, but that the rent is too low. There is much
talk about agricultural credit. It seems to be overlooked
that landowners are lending the capital value of their

[1] "The Making of the Land in England," Albert Pell, *Journal
of the Royal Agricultural Society*, 1887, 2nd Series, vol. XXIII,
p. 355, and 1899, 3rd Series, vol. X, p. 136.

"The Rent of Agricultural Land," R. J. Thompson, *Journal of
the Royal Statistical Society*, vol. LXX, 1907, p. 609.

holdings to their tenants at less than the market rate of interest. No Government could do more, and, with proper regard for the Exchequer, no Government ought to do as much.

English land is cheap both to buy and to rent. The average pre-war value of farm land in Belgium ranged from £54 to £59 an acre, and the average rent was 36s.[1] It should be noticed that, even in that country of small owners and small tenants, there is the same excess freehold value compared with rent, of which our critics complain in England. To pass to very different conditions—the average post-war value of all land under crops in New South Wales is £120 and in Victoria £100 an acre[2].

These figures may at all events serve to bring into a sense of proportion the widespread prejudice that exists against the landowner. Much of this prejudice is due to a confusion between urban and rural conditions. Our predominantly urban people confuse the country squire with the owner of town ground rents or house property. As so often happens, an understanding of the cold facts would serve to dissipate a great deal of hot feeling.

There is more in the question of amenity value than appears at first sight. It is really an important factor in past and present agricultural economics. Without it less capital than at present would be available. There are more profitable investments for both landowner and tenant, and, if there were no amenities to attract them, the land would really become under-capitalized. Farm

[1] Seebohm Rowntree, *Life and Labour in Belgium*, 1910.
[2] Sir Frank Fox, *Spectator*, 6 November, 1926.

buildings would not be maintained or replaced on their present high level, and much more arable land would go down to grass. Because of the amenity value of their estates, landowners have been willing to purchase them, to maintain and improve the houses, cottages, farm buildings and other equipment of the land, and have been content with a very small interest on their money, not merely on the purchase price, as our reformers allow, but on that expended afterwards and continually in improvements.

In regions of rich soil, such as some parts of the Fens, agriculture can succeed without amenity value, but poor land, such as much East Anglian clay, goes out of cultivation in bad times sooner in those regions where amenity values are low. Moreover, it is generally admitted that cottages and other buildings are better on the larger estates, that is, where amenity values are high; it is the small man who usually lets them get out of repair.

Amenity value is a real asset to the private landowner, and he is prepared to pay for it by providing and maintaining the fixed capital of the land below the market rate of interest—whereby agriculture and, through it, the nation gain. On the other hand, this form of amenity has no value to the State or to a County Agricultural Authority. Hence arises a dilemma which appears in all schemes for nationalizing agricultural land. Either amenity value must be destroyed, which is unjust to its owners and tends to injure agriculture and reduce national wealth, or else, by paying for it, the nation is saddled with a scheme even more unsound financially than that of the Liberal Land Report. Personally, if we are to have nationalization, I think part of the amenity value should be paid for by the State, which will be responsible for the change, and part might be charged to the cultivating tenant, who will, at all events, retain some of its elements. But he would then be little better off than if he bought his farm now from his landlord; his dislike to schemes of nationalization would be intensified.

Even if carried out gradually and with adequate compensation, all schemes of nationalization tend towards the eventual public ownership of the land, and the extinction of the class of country landowners.

To sever finally the connection with their ancestral acres of our few remaining mediaeval Houses, and of the many less ancient but still historic names, may not shock an unromantic age. But to drive all landowning families away would have a direct economic and social effect on the countryside. At present, with the squire and the parson, there are often two men and their families in a village with a somewhat wider experience and a more detached outlook than the bulk of the inhabitants, and ready with help or advice for their neighbours. How would it be if all who could afford it migrated to the residential towns? I do not think that the labourer would gain in the long run by being left face to face with the farmer alone on the countryside. Landowners are even now being supplanted on County Councils and other local bodies by farmers whose chief desire is to keep down the rates, and this tendency must inevitably be hastened by schemes for expropriation[1].

[1] See a letter from Mr Henry Hobhouse in *The Times* of September 19th, 1925.

Already Labour orators are saying openly that the new farmer is worse than the old squire[1].

Probably the irresponsible Labour orator wishes to oust the farmer also, and give all power to the labourers or their trade union organizers. But the official Labour pamphlet has the wisdom to recognize that, for any time which can be foreseen, tenant farmers will hold from the State on leases or agreements similar to their present ones. The removal of the landowner would leave the farmer class firmly established as the head of the village community.

Lest it be thought that we exaggerate the benefits to the countryside of resident landowners and the losses that their removal would entail, let us quote from some who wish to expropriate them. There is one authority which perhaps even the authors of *The Land and the Nation* will accept—that of Mr Lloyd George, who said in the House of Commons on July 16th, 1926:

> He had never attacked the good landlord in the country. Such a landlord's estate was, in his opinion, the model on which smallholdings ought to be founded by the County Councils. There was security of tenure, rent was not raised on the tenant's improvements, and money was spent on repairs and drainage...the model was not ownership but tenancy on a good sound estate.

Verily Saul is among the prophets, and, with this verdict ringing in our gratified ears, we might well leave our *apologia* for the country landowner.

But let us also turn to the book of Mr Orwin and Colonel Peel, who, as they themselves write, advocate the acquisition of the land by the State "from motives

---

[1] *Cambridge Daily News*, November 12th, 1925.

diametrically opposed" to those so often advanced by land reformers. Orwin and Peel point out that the forced sale of estates first to speculators and then to the tenants drives away the landowner.

As regards rural society generally, the disappearance of the landowner deprives it of its natural focus. With the dispersal of his property and the sale of his home nothing remains to tie the squire to the locality in which he has exercised for so long in greater or less degree functions of leadership and wise control. At the present time hundreds of country houses, once the centres of active social life, are in the market for disposal... many people do not realize the extent of the collapse in rural society which is the first result of their abandonment, bringing discomfort and even misery to many of their more humble neighbours. Indeed, there are some who think that it is not for the good of rural society that the landlord element should be eliminated from it by the uncontrolled operation of economic pressure. The greatest single cause of social unrest is the disintegration of classes. Where everything is understood everything is forgiven, and people can only understand each other when they have opportunities of mixing freely one with another; on the estate and the farm, at work and at play, all classes on the land are in almost daily contact.

It is to prevent this unordered breakdown of rural society that Orwin and Peel propose their remedy. The State is to buy the land. The tenants are to keep their holdings under the direction of district and county land agents, and the landowner, in the secure possession of a trustee security in the form of Land Stock, is to take his house from the State on lease and remain a leader in village life and an influence for good as he has been in the past.

Now this is very different from the objects of the Liberal and Labour reformers, who wish to eliminate the landowning class altogether. It needs careful and sympathetic attention. I have little objection to the principle of Orwin and Peel's scheme from the owner's standpoint. My fear is that it would not produce the effects they hope on rural society. Doubtless for a time landowning families would continue their work, but the feeling of responsibility must inevitably be weakened, and there will come a generation which loves not country life. In "good residential neighbourhoods" the descendants of the old landowners may remain permanently, but, in duller regions, where the ownership of the soil now holds them, they will soon drift away, and it is there they are specially needed.

If we are to have nationalization, let it be carried through in the spirit of Orwin and Peel, and, with modifications, on their lines. But I am not yet convinced that the game is up, and as long as landowners will carry on, or others come forward to replace those who fall by the way, I think the State gains by leaving things alone. The pleasure, the interest and the opportunity for social service which the possession of land gives to the private owner are used by the nation under the present system to obtain from him capital below the normal rate and, in most cases, good management also. The landowner gets what he wants in well-earned satisfaction, the tenant gains, and the nation benefits. Why disturb what all unprejudiced observers, and some even who are prejudiced against the landowner, agree is an admirable arrangement?

Another argument for nationalization, used even by Orwin and Peel, is also met by the facts brought out in this chapter. It is often said that the State is precluded

from helping agriculture because any increased prosperity thus given to the farmer would pass by way of
increased rent to the landowner.

The reply is as follows. Firstly, with Wages Committees keenly on the look-out, it is certain that the
first increase in profits would be absorbed in higher
wages long before rents, which always lag slowly behind
profits, had time to be adjusted. This is but right; the
labourer has first claim on any increase of prosperity.
Secondly, if agricultural wages were raised to conformity
with those in other equally skilled trades, and a surplus
remained which, as tenancies changed, led to some increase in average rents, one must ask—Why not? Partly
by inevitable economic accident, partly by deliberate
monetary policy and social legislation, the nation has
thrown an additional contribution towards the costs of
the war and of the peace on the agricultural landowner,
and reduced by 40 per cent. his pre-war real rents, as is
proved earlier in this chapter. If by State action some
of that burden is raised, the nation is only repaying the
landowner a part of the debt it owes him.

That is the answer which should be used by those who
advocate protection or subsidy. But the responsible
Associations of landowners and of farmers ask for neither
of these State contributions, unless the nation desires
to change the natural economic trend of farming and
increase artificially arable cultivation. If left alone, they
are prepared to make the best of things and adjust agricultural methods to the economic conditions of the time.

The case for the landowner as presented in this
chapter shows that he has an incontestable moral claim
to share in any increased prosperity of agriculture, how-

W                                                        9

ever it may come, even by State assistance. But other reasons against protection or subsidy remain, and personally I would rather try to develop orderly marketing and the stabilization of prices, and otherwise leave recovery to less dangerous, if less certain, causes.

Will those causes, if left to work themselves out unaided, bring recovery? It must not be forgotten that present real rents are little more than half those needed to put landowning on a sound economic footing, and enable landowners to maintain the equipment of the land on a pre-war level, without the present heavy drain on their private incomes. Probably post-war prices will settle down at about 150 compared with a pre-war 100. To regain equilibrium, therefore, costs of building and repairs must fall from their present level of perhaps 200 to the normal 150, and rents rise from perhaps 120 to that same 150. Till that happens, the landowner is paying not only his fair share in income tax, but an unfair additional share as well, towards the cost of the war and the exactions of the peace.

The general agricultural outlook will be considered in a later chapter. I think that the balance of probability is that things will slowly right themselves. And, even if real rents rise to their true relative level, the gain will not all go to the privy purse of the landowner. Part of it will benefit those dwelling on his land. For, as Mr Orwin and Colonel Peel say:

> The net income is rarely available for the landowner to the extent that the incomes of other investors are, for the owners of agricultural property have behind them a tradition of sharing their possessions with the community in which they live to an extent unknown of any other class.

Chapter VIII

## THE MANAGEMENT OF AGRICUL-
## TURAL LAND BY THE STATE

In considering the probable results of nationalizing agricultural land, nothing is more important than a study of the various methods of management which have been proposed.

Mr Orwin and Colonel Peel advocate a simple and business-like scheme. Properly qualified land agents, each managing 30,000 acres more or less, according to the intensity of agriculture, would be given responsibility for prompt decisions under the general supervision of County Land Agents. The County Agents would be appointed by and would report to a Chief Commissioner of Lands in the Ministry of Agriculture, but no interference from Whitehall in the details of actual management would be allowed. The County Land Agent would be *ex officio* a member of the County Council Agricultural Committee, and thus would be in touch with the work of agricultural instruction, advice, etc. But apparently the County Council would have none of the powers of an owner. These powers in effect would be vested in the County Land Agent as Trustee for the State. Thus the scheme is one of pure bureaucracy, and is subject to all the objections, both real and imaginary, against management by Government officials. But the agents would be well-trained professional men, and the service should prove attractive enough to secure a good class of applicant. The scheme would work.

The Labour pamphlet states:

Under our proposals the present County Agricultural Committee would be re-constituted, and they would enjoy wider powers. Each Committee would, we suggest, consist of (a) an equal number of representatives of farmers and farm workers[1], chosen by their respective organizations, and (b) members appointed by the Ministry of Agriculture from a list including experienced persons suggested by the County Council and other appropriate bodies, including small-holders, this section to constitute a majority of the Committee.

All questions relating to agriculture in the county would come within the purview of the County Committee. It would enjoy powers similar to those authorized by the Agriculture Act, 1920 (now repealed), for the purpose of maintaining a good standard of husbandry. It would be empowered to enforce good husbandry, to improve existing methods of cultivation (e.g. in regard to the eradication of weeds and the adoption of suitable methods of manuring), to provide necessary works of maintenance (e.g. the cleaning of drains, embankments and ditches, the repair of fences, etc.). In the case of grossly mismanaged farms, there should be power to dispossess the holders....

The Committee would therefore be a very important body, and, in order that the Committee shall not be confined to the well-to-do, the out-of-pocket expenses and loss of wages incurred in attending meetings should be met out of public funds.

[1] In parenthesis, may I protest against the growing fashion of using the name "worker" with no qualification to mean wage-earning manual labourer only? Are not farmers, factory managers, professional men and most landowners "workers" also? The average smallholder works, even with his hands, harder than the average labourer for wages, and besides has the anxiety of managing his own business. The restricted use of the word "worker" has an evil psychological effect when reiterated as it is by those wishing to create ill-feeling and class antagonism. The manual labourers come to believe that they are indeed the only workers, and cease to realize the fact that without direction and management manual labour can do but little.

The present County Agricultural Committees are Committees of the County Councils. But the connection of the Committees here proposed with the County Councils is shadowy. They are in effect new *ad hoc* authorities. This brings the general plan of the Labour scheme very near to that of the Liberals, though details are different, and the Liberals add to the Committees a large element directly elected by the agriculturalists of the county. We can therefore describe the Liberal scheme and then consider the probable effect of the Liberal and Labour schemes together.

It is remarkable that the authors of the Green Book summarily reject the idea of entrusting land to the County Councils. They say:

> In the course of our enquiry meetings we found that the vote of village audiences was unanimous against handing over the administration of Cultivating Tenure to the County Councils. So strong was this feeling that, even when our speakers suggested ways and means of making the County Councils more thoroughly representative of the popular will, the village audiences refused to be persuaded.

I think this point is of some general interest. The Liberal reporters assume that the adverse opinion of village audiences prove County Councils to be a failure, and perhaps wonder that other people are not satisfied with the usual Liberal remedy of amending the constitution of the Councils so as to make them "more thoroughly representative of the popular will."

I regard the opinion of the village audiences rather as another illustration of the disillusionment which is now so common when people are faced with the disagreeable fact that even the democratic institutions they have obtained will not overcome the hard and unalterable conditions of life.

But that, as Kipling used to say, is another story, about which we shall have something to write below. *Revenons à nos moutons* and to the proposed County Agricultural Authority which is to be set up to control them.

Disillusioned with County Councils, village audiences demanded something which would bring the millenium more quickly, and the authors of the Green Book set forth valiantly to satisfy them.

They lay down the following principles:

> The administrative body should be popularly elected on the widest possible franchise. Each administrative area should have its own representative body. Each County should be autonomous to the largest degree compatible with safeguarding the interest of the nation as a whole....Rural areas must be rescued from urban preponderance.

On these lines an *ad hoc* Committee is to be constituted. Half the members are to be nominated and half provided by rural districts and by those urban districts which contain a minimum of 500 agriculturalists each. The members from the rural areas are to be elected directly on the County Council franchise, and those from the urban districts appointed by the Borough or District Councils.

*The Land and the Nation* itself admits that this reversion to *ad hoc* authorities is contrary to the policy of recent years, which has tended steadily to the concentration of local government in the hands of Municipal and County Councils. To ask the ordinary elector to vote once for Parliament and once for his Local Authority

seems to be to ask as much or more than he will give.
The poll at County Council elections is sadly small. But
will the elections to the County Agricultural Authority
create more interest? Public work of any sort demands
time. Attendance at the county town is an expensive
business for persons living miles deep in the country.
And therefore it is proposed to pay members' expenses,
including payment for lost time. I fear that even this
provision will not secure the services of the best type of
agricultural labourer, who is a busy man. It is more
probable that trade union organizers will become can-
didates. I fear, too, when the inevitable discovery is
made that democratic County Agricultural Authorities
can no more bring prosperity to agriculture on a falling
market than can private landowners, or enable farmers
to pay high wages out of their losses any better than
existing Wages Committees, that interest in the elections
will fall to an even lower level than in those for County
Councils.

Of the other half of the Agricultural Authority, con-
sisting of nominated members, two-thirds are to be
nominated by the County Council. To this no exception
can be taken. The other third is to be nominated by
the Ministry of Agriculture, but only after consultation
with "the National Farmers' Union, the Workers'
Union and other recognized agricultural organizations
in the County." If this consultation is to be more than
a form, it seems likely to increase the difficulties of
selection of the Ministry, and not necessarily to lead to
the best appointments.

The Liberal Conference, perhaps wisely, did not tie
itself down to approval of the details of this scheme,

but confined itself to resolving that "there should be in every county a representative Agricultural Authority including owners"—this last word being an afterthought added by amendment—"farmers, smallholders, allotment holders, and land workers . . . advised by responsible officers of high standing."

For an effective criticism of the proposed County Agricultural Authorities, we need only read the speeches at the Liberal Conference, where strong objections were raised to the establishment of new Local Authorities independent of the County Councils. It was also pointed out that, under the conditions of *ad hoc* election, "men will be elected . . . for all sorts of reasons which have nothing to do with agriculture."

There is much truth in this criticism. Besides suffering from the innate faults of their own constitution, the proposed County Agricultural Authorities must be troubled by duplication of duties and conflict of authority with the County Councils, which are still to retain functions both directly and indirectly agricultural.

It is interesting to note that the idea of popular election is not taken up in the pamphlet which describes the Labour policy. The Committee therein proposed is made up of representatives of unions and a majority of nominated members. The provision for direct representation of farmers' and labourers' unions is as characteristic of "Labour" as direct election is of Liberal politics, for the constitution of the Labour Party is based chiefly on the economic organizations of the manual labourers. The scheme it proposes is dangerous, because, while trade union organizers may be effective on Wages Committees, they are not likely to be always useful members of bodies that have to deal with technical agricultural questions. Moreover, representatives of unions of employers and employed seem unable to avoid assuming an antagonistic attitude, even though as individuals they might work together amicably. I believe that on this rock the efficiency of all such schemes would be wrecked.

The duties of the new body, however it be constituted, are to be both heavy and responsible. It is wisely proposed to carry out the actual transfer of the land, the fixing of the landowner's annuity, etc., by means of temporary Commissioners, as that work is judicial rather than administrative. But once the scheme is started, the County Authority will have control. Whether it assumes possession of the whole of the agricultural land of the county at one swoop, as advocated by the Green Book and the Labour Party, or gradually takes it over as it comes into the market or is specially needed, as agreed by the Conference of the Liberal Federation, it will become a very great landowner, and, if it is to do its work conscientiously, must take to heart as many of the landowner's anxieties and responsibilities as lie within its power. It cannot hope for the personal touch, which means so much on a good private estate, but it must take over all the many points of detailed business which pass through a private estate office.

Its wider duties are set forth in the Green Book and the Labour pamphlet and at greater length in the *Report* of the Liberal Conference. According to the latter, it has to see that all land is well cultivated; meet the demand of qualified applicants for smallholdings or family farms; ensure that every labourer who desires it shall obtain at least half an acre of land at a fair rent and provide allotments where required; take part in the administration of both long- and short-term agricultural credit; promote co-operation, more efficient marketing, agricultural research and education, the development of village industries and the amenities of village life.

It will be seen that the members of the Conference expect the County Agricultural Authority to carry on many and varied functions. One wonders if the Conference realized the mass of detail involved in the mere commonplace management of such great areas of land, and one doubts whether the Authorities will have much time left to devote to the broader problems of policy mentioned above.

However that may be, a large scheme of rural development is outlined in the Green Book, in the *Report* of the Conference and in the Labour pamphlet. But most of the objects aimed at are common to all agricultural policies and to all political parties. They can be secured with no change in the system of land tenure. A living wage, good cottages, ample gardens (or failing them allotments) for those who want them, an increase in the number of smallholdings, easy credit, are found in all manifestos, even in the Government White Paper, and the modest statements of the latter have the great advantage that they are, even now, in process of being carried out by legislation.

On the points where the Liberal or Labour pronouncements differ from others a few words may be added.

(*a*) According to the Green Book, a living wage, where not at present payable out of the profits of a farm, is to be made possible by a forced reduction in rent. At present this object is secured in a more natural way by the action of Wages Committees. If a farmer cannot pay the wages fixed, he naturally falls back on the tenant farmer's first line of defence—an appeal for a rebate or reduction in rent. To review in some judicial or pseudo-judicial manner, the rent of each individual farm, as is apparently contemplated, opens out many difficulties. Who is to say, for example, whether the farmer is suffering from an excessive rent, from unavoidable economic causes, or because of his own want of skill? It is claimed that landowners are even now unable to maintain the equipment of their land because of poverty, due to insufficient returns therefrom. This at all events shows that the limit is nearly reached. Any further reduction, especially an artificial reduction, of rent must intensify the evils that are envisaged. After all, agriculture resembles other industries, and, if the net returns, including amenities, etc., are not adequate, as long as any private landowners are left, capital will cease to flow into the business.

When all land is owned by the State, other dangers are considerable. Wages can be raised if rents are reduced, and intense political pressure will be exerted to increase the labourer's remuneration at the cost of what is regarded as the bottomless purse of the taxpayer. That is the road which leads to national ruin.

(b) No explanation is given as to how security of tenure for existing farmers is to be reconciled with a large extension of smallholdings and family farms, and a right to at least half an acre of land for each labourer.

(c) The Liberal obsession about security of tenure has been dealt with in Chapter IV. The right to improve and

compensation for improvement are already given to
tenants by the Agriculture Acts. "Fair rents representing only the agricultural value of their land" are
and have been the universal rule. Indeed a great part
of the Green Book is given up to a complaint that the
selling price of land is artificially swollen by monopoly
and amenity values compared with the capitalized value
of the rent, as representing the fair agricultural value.
In one place the authors allow to escape them a quota-
tion from Sir Daniel Hall that "rents are below their
true economic value in England." They use this fact to
claim the "excess of the real over the rental value of
the land" for the State. But apparently they do not
see how completely it invalidates their general picture
of a suffering tenancy in the pitiless though failing grasp
of "landlordism."

(*d*) The Labour pamphlet proposes that all game
should be the property of the occupier, who should be
responsible for damage done on neighbouring holdings.
"Drastic amendments to the law" were demanded by
the Liberal Conference to give cultivators complete protection from damage by game and foxes. Perhaps some
of the readers of the *Report* of that Conference, possibly
even some who attended it, never heard of the Ground
Game Act of 1880, whereby every tenant can kill hares
and rabbits, or the present very stringent provisions
for compensation for damage by other game. Pheasants
may be overdone in some parts of the Eastern Counties,
but much of the harm put down to pheasants is due to
wood-pigeons, most of them migratory. The *Report*
admits that "partridges do no damage to the crops,
and in fact, consume large quantities of harmful grubs."

Game usually has a local habitation and therefore an owner, who can, perhaps, be proceeded against where damage is excessive. But how the law can give "complete protection from damage by foxes" I fail to see. Who is to be responsible? Would an action at law, or an injunction by the County Agricultural Authority, lie against the fox? I agree there is often heavy loss, insufficiently met from the hunt poultry fund. But, after all, the preservation of foxes is voluntary, and is done by the occupiers of land, not by the hunt. It is fair to assume therefore that the majority of farmers in hunting countries think the sport and the business that it brings are worth the cost. Indeed, most country folk enjoy hunting on horse or on foot, or gain by the money it brings into the neighbourhood. One thing is certain—fox-hunting would not survive for a year if the general sense of the countryside were not in its favour.

But of course the chief function of the new Committee will be to secure good cultivation from its tenants. As the Green Book puts it:

The County Authority's duty—its primary and its ultimate duty—that of securing good cultivation of all cultivable land in its area, will necessitate bringing all farmed land under systematic survey. This survey will be carried out by the Authority's officers, whose work will be similar to that of land agents employed now on large estates. It will be the duty of the C.A.A.'s Officer to report any holding which apparently is not being satisfactorily used. It will not be his duty nor will it be the duty of any officer or member of the Authority, or any other person, to dictate to the farmer the agricultural methods he should pursue. The standard of husbandry will be judged by the circumstances of the district.

Now I fully agree that it is desirable to bring all

possible pressure to bear on bad farmers, and that, in present conditions, it is difficult for private landowners to do so. But that is because the Agriculture Acts have made it practically impossible to get rid of a tenant, save at heavy cost. It would be easy to devise effective methods of meeting this difficulty without upsetting the whole system of land tenure in England, which, again to take a text from Sir Daniel Hall, "still seems the most effective form of dealing with the land on a large scale."

It seems very unlikely that the new Agricultural Authorities will bring to pass the improvement in the standard of agriculture expected. As Orwin and Peel say, "'Orders' to farm are quite unworkable." For more drastic action, if the new Authorities resemble the present Committees, there will be the same difficulty in getting them to declare their neighbour's land badly farmed, and thus to deprive him of his livelihood. Indeed, as the initiative must come from them, the difficulty will probably be greater than it is now. If their officers have, or take, greater power, and if those officers happen to be competent and energetic men, doubtless the efficiency of the present method of removing bad tenants will be increased. But then all the usual feelings, justified and unjustified, against "officials" will be invoked, and many obstacles put in their way.

If, on the other hand, these powers are circumscribed, and really democratic Authorities take charge, I regard the prospect as far worse. In fact, anyone with some knowledge of farming on the one hand, and of the working of Local Authorities on the other, will feel the inherent difficulties of controlling all the land of a county, and maintaining its standard of cultivation, by means of a Committee—any Committee, but more especially such a Committee as the Liberals or Labour men contemplate.

In the Liberal scheme, the chief point of instituting *ad hoc* Authorities is to secure complete democratic representation confined to agriculturalists. That is to say, as far as the elected half of the Authority is concerned, owing to their greater numbers, labourers and smallholders will completely outvote the larger farmers and the residual landowners. What will happen can be foretold from the account of village meetings given in the Green Book itself[1]: "Our Enquiry Campaign meetings prove clearly that there is a strong demand for limiting the number of acres which any one man may farm."

The members of the Liberal Land Committee themselves are plainly frightened by this discovery. They say:

We take the view that in agriculture, as in other industries, there should be rewards for enterprise and scope for ambition....Some land...may in the future be farmed most satisfactorily from every point of view in large scale mass-production units....Any hard and fast limitation of acreage would make such use impossible.

But, having recognized the danger, the Committee has no safeguard to suggest, save that "no general ruling shall be made limiting the size of a holding in Cultivating Tenure"; in specific cases the County Agricultural Authorities are to be left free to limit it as they will.

The existence of a demand for limitation is natural enough to the peasant mind; yet it is clearly against the

[1] Page 343.

good of agriculture and ultimately against the good of the community. Corn growing can only be carried on economically on the large scale. To rise to be foreman and then manager on a large farm is an alternative, and sometimes a better, method of advance than the painful road of a smallholding. But inevitably limitation would be put in force if the principle of democratic representation were applied in this *ad hoc* manner. The essence of the constitutional system as it has grown up in Great Britain is representation by localities, which, save in instances too rare to produce much effect, contain inhabitants of all sorts. Thus, speaking broadly, the interests of the community tend to outweigh the interests of one industry, as judged from the narrow outlook of that section of it which chances to be greatest in number. And it is this historical accident which has probably made British democracy a success, or, if a change of emphasis be preferred, has prevented it hitherto from doing much harm. County Councils are elected in this well-tried manner, and for the most part, save for turning out bad farmers, for which they are unfitted, they and their Agricultural Committees do their work efficiently and well.

*Ad hoc* authorities have been tried before, but they have dealt with functions like education or poor-relief which cut across industrial divisions. No attempt has been made to control specific industries in this way, and the dangers are manifest. The hampering effect of a really democratic County Committee, dominated by the peasant outlook, on the enterprise of large and successful farmers would probably diminish the general efficiency of the industry more than all the efforts of their executive officer to improve the standard of bad small cultivators would do to increase it. Large farmers might well be driven out of business, and the energetic, able, and skilful agriculturalist is, or soon should become, a large farmer. To lose such men would be a national disaster. To give their brains and their capital free play is the best hope for arable farming.

## CHAPTER IX

## THE FINANCE OF NATIONALIZATION

W E have already considered the finance of the scheme for nationalizing agricultural land under the head of the farmer's security of tenure, and also in another chapter from the point of view of existing landowners, and acknowledged the improvement effected in the Liberal proposals by the Conference of the National Liberal Federation, who fully recognized the gross injustice to present owners of the scheme set forth in the Green Book. We have now to study both this same scheme of finance and other schemes of nationalization which are before the nation as they affect the National Exchequer.

Under the Liberal scheme, as modified by the Conference, the nation will take over land, as it is offered for sale, at a price fixed, or to be fixed, by the Valuation Department, deductions being made for any "monopoly value," whatever that may mean, and for any adjustment of rent necessary to enable tenants to pay their men a living wage. The sense of the Conference seemed to be in favour of paying something less than the present selling value of the land, though something more than the Green Book provided. The landowner will be fortunate if he gets what the Conference intended, and the nation will be more honest, though poorer, than the authors of the Green Book or than those of the Labour pamphlet hope. But, to put on the matter a light the most favourable to the Exchequer, let us take the original scheme of the Green Book, whereby the landowner is to receive an annuity equal to "the fair net rent receivable." The book says:

> The reason for proposing an annuity payment is that the urgent need of agriculture is a sufficiency of capital; that the capital improvement of its land is the soundest investment the State can make, and that the State's credit resources should therefore be used to assist the cultivator rather than to pay a lump sum for the purchase of the freehold.

Do the authors of the Green Book really suppose that a national transaction of this magnitude would be carried out by paying cash for the freehold? Do they imagine that the creation of perpetual annuities to the amount of some £48,000,000 a year would have no effect on the State's credit[1]? Such a creation comes to precisely the same thing as the issue of the capitalized value of these annuities—say £960,000,000—of 5 per cent. Land Stock, which would be the natural way of carrying out the "purchase of the freehold."

Let us imagine then that the transfer is effected. The State owns the land, and, whether the amount is called "annuities" or "interest on Land Stock," has to pay in perpetuity a sum of £48,000,000 a year and the costs of administration. Will "the fair net rents" recoup the expenditure? Let us examine from this point of view the various proposals which have been put forward.

On the scheme of the Green Book, net rents and costs are adjusted so as to amount to the same total, and the

---

[1] £48,000,000 a year was estimated as the approximate net value in 1913 of farms, farmhouses and farm buildings in Great Britain (see Sir J. C. Stamp, *British Incomes and Property*). The total national income at that time was generally supposed to be something over £2,000,000,000. The national income must now be above £3,000,000,000 (in post-war pounds), while nominal agricultural rents have increased by only about 20 per cent.

accounts nominally balance. But more remains behind. The authors regard the usual 5 per cent. paid to land agents by landowners as an excessive charge for management, and propose to debit the "cultivating tenants" with 3 per cent. on the net rents only, expecting it to cover the cost.

Now on this estimate there is much to be said. The private landowner's 5 per cent. is reckoned on the gross rents, and, moreover, only covers ordinary management and rent collecting. There are always extra charges for agent's out-of-pocket expenses, and occasional professional fees for special services on a change of tenancy. That 5 per cent. is not an excessive price to pay on ordinary small estates is shown by the fact, quoted at the Conference of the Liberal Federation, that even with the very large property of the Crown Estates—70,000 acres of agricultural land—and very efficient personal administration, management charges amount to 4 per cent. of the rental.

It is improbable that the County Agricultural Authorities would work as cheaply as the Crown Estates Office. County Councils spend much more on their small-holdings. The only one for which I have details at hand spent on management in the year 1924–5 a sum which works out at 20·9 per cent. of the gross rents. Doubtless larger farms would cost less, but it must not be forgotten that it is an essential part of the scheme to pay the many members of the County Authority their travelling expenses and a fee for lost time. Other administrative expenses would be sure to arise, bringing the minimum cost much above that usually paid by a private owner. We may safely expect the real cost of management to fall somewhere between the 4 per cent. of the Crown Office or the 5 per cent. of private estates and the 20 per cent. of the County Council smallholdings.

Having hopefully fixed a charge on the tenants of 8 per cent. for management, how do the authors of *The Land and the Nation* propose to meet any excess? Apparently it is to come out of a "Central Land Fund" accumulated mainly from the proceeds of the sales of land for building and other non-agricultural purposes[1]. That is to say, the authors propose to use capital to pay for deficiencies in income.

We have already, in considering security of tenure, pointed out the one-sided nature of the arrangement whereby a cultivating tenant enjoys a fixed rent in good times, and can throw up his holding, and thus get the rent lowered, when times are bad. This defect in the scheme is even more important than the under-estimate of the cost of management.

The whole idea of giving State tenants fixity of tenure was severely criticized at the Liberal Federation by Mr Geoffrey Howard, who said:

Is this proposal a Liberal proposal at all? As I have known Liberalism it is not. Cultivating tenure proposes to create a new hereditary class with a vested interest. You are proposing to give the present tenant an hereditary right to sit upon his farm at a fixed rent. You are proposing to give the landlord a fixed annuity. Every Liberal wishes to see the farmer assured of the result of his own energy and his work, but everyone who has studied this rural question knows that profits alter not merely upon the individual work of the farmer but as a result of lower prices. If there is a rising world price, under cultivating tenure the individual will get

[1] *The Land and the Nation*, p. 351.

the benefit. If you get a fall in world prices, which is equally likely, your man under cultivating tenure can get out of his holding at a year's notice and throw his holding back upon the community, who will have to let it at a lower rent. You let a farm at a price it will fetch, according to the price of the commodity grown upon it at that time. If there is a fall in the world's prices the farm will have to be let at a lower rent. You have guaranteed the landlord his annuity; who is going to fill in the gap? The taxpayer.

One other point. The Liberal policy, which I have fought for in the House of Commons and which I shall continue to fight for as a Liberal, is the encouragement of the small man on the land. I say to my audience, which contains many men who have worked on County Council Smallholdings Committees, that the problem of smallholdings is not a question of land. Land can be got, but the problem is to put up the steading at a price which can be paid for.

Exactly. It is not the landowner but the builder and his men who are the profiteers.

We must also remember that the Liberal Conference, admitting the injustice of the Green Book scheme, adopted amendments which have the effect of increasing considerably the compensation to be paid to land-owners. Apparently no increase is to be made in the rents. Hence, if anything approaching justice is to be done, the nominal balance between the annual sum paid by the State on annuities or Land Stock, and the "fair net rents" it will receive is upset, and the result is still more costly to the National Exchequer.

The Labour policy goes less into details; consequently its effect is less clear. It contemplates expropriation on the basis of the Schedule A assessment, less deductions for needed repairs and improvements. This may, perhaps, come to somewhat the same as the Liberal scheme. But "Labour" seems also to expect the proceeds of the rents to pay a sinking fund to extinguish the Land Stock, as well as to defray management expenses. Thus either there will again be an annual deficit, or the interest on the Land Stock must be lowered so as, in effect, to throw the burden on the expropriated owner.

Mr Orwin and Colonel Peel avoid all the expense of committee management. Their County and District Land Agents, if left free to function as is the Agent for the Crown Lands, might get the cost of management down to some figure in the neighbourhood of 5 per cent.

But even this last, more feasible, scheme has to face the usual dilemma. Either the amenity values, which to a private owner are an essential part of the property, have to be ignored and so mostly destroyed, which is unjust to the owner, or, by paying for them, the State acquires an investment which will not pay its way financially. As the State will be responsible for the destruction, and since such amenity values as survive will be shared between the nation at large and the tenants, I think that just compensation for them should be paid by the State, and part recovered in the rent if it prove possible.

Unless there is a considerable rise in the general level of agricultural prices, produced by monetary changes or by a growing world shortage of food, it seems certain that any scheme of nationalizing agricultural land which is just to the landowner will involve a financial loss to the State. If all land were taken over and a real monopoly established, the State might recover this loss on the profits of building land near towns. But that would be a speculative investment, dependent for success on continually expanding urban requirements.

# PART III

## *THE FUTURE OF RURAL ENGLAND*

### CHAPTER X

## OUR PRESENT DISCONTENTS

In considering the future of rural England, it is well to deal first with the conditions which the country shares with the town, conditions common to the whole nation, before passing to those special to agriculture and other rural industries.

For example, the most important and insistent point of all, the low wages of the farm labourer compared with the earnings of men equally skilled in other walks of life, cannot be understood by a study of agriculture alone. On the average, farmers' profits are barely enough to pay present wages. Real rents have fallen till it is a doubtful point if capital can be found to maintain the permanent equipment of the land: no further raid can be made on the landowner. It is high costs of production that are the trouble. Local services, transport, distribution, and especially building repairs, cost more than the rise in the general price level warrants. The municipal servant, the railway man, the shop employee, the building operative, in present economic conditions, are obtaining part of their high wages at the expense of the agricultural labourer, and at the cost of unemployment in other unsheltered industries. The wages are not too high for the needs of the men; but they are higher in proportion to production than the economic state of the country can bear for the moment. Either a considerable increase in output per man, or a temporary reduction in nominal wages in sheltered trades is necessary to allow unsheltered wages to rise, and to enable the nation to recover the strain of the war and start afresh. To this point we shall return.

The example may serve to show that we cannot consider the future of agriculture in isolation, and to explain why this Part of the book begins with some general thoughts on our present discontents.

The desire for some radical reconstruction of rural life, when it is not merely a misunderstanding of the causes of agricultural depression, is usually one symptom of the present general disillusionment with life and not least with all systems of Government. Both those who have an excessive faith in the principle of representative institutions, and those who regard it as an outworn dogma, and look either to Italy or to Russia as a road to Utopia, alike ignore the basic facts of existence.

Here are we, teeming multitudes, all of us with limited faculties and most of us very stupid, placed in a mysterious world we do not understand, in natural conditions generally hostile, from which somehow we have to wring a living. We have been men only for a brief time, since our ancestors came down from their trees perhaps some three million years ago. We have been in any sense civilized for only about five thousand years, and that with periodic lapses into barbarism, confusion and poverty.

We have gained a limited control over our physical

environment, and developed a social and economic system which, in a wasteful but hitherto effective way, provides for the accumulation of capital and thus performs the miracle of housing, clothing and feeding large and growing populations, and giving them in parts of Europe, and in the homes overseas of its daughter nations, a standard of life far higher than such numbers ever had before. The system is imperfect—of course it is imperfect—but no other system has done so much for the mass of the people in the past, and there is not a scintilla of evidence that any radically different system will do as much in the future. Russia, who thinks she has tried a different system, is living on the accumulated capital in buildings, material and technical skill inherited from the old *régime*, and on the exploitation of the peasants, who have reverted to an individualist economy—even thus living on a lowered standard. What will happen as the fixed capital finally wears out remains to be seen—probably further tribute levied on the peasants or a complete breakdown.

As regards Government, it is a difficult job at the best. No ideal system of government ever has been devised, and it is safe to say that none ever will be. Our own muddle-headed, illogical, patchwork constitution seems to have worked better than any other known arrangement; but even that is a faulty thing, and seems clearly to be past its best.

As long as power was confined to the few, the many could blame them for what went wrong. Now that all have power, they are surprised and angry to find that improvement still comes slowly, for a time perhaps not at all. They grow suspicious, and persuade each other that they are the victims of a conspiracy. The enormous gain in the standard of life of the wage-earners is easily forgotten, and anyhow is less evident than the inequalities that remain. They do not realize that in economic advance Governments can do little good, though much harm; that man is still much the same; and that nature is still regardless of his desires or his welfare.

The Liberal hopes to mend things by broadening still further the basis of democracy, formerly in Parliament and now by instituting democratic County Agricultural Authorities. The Conservative and the Socialist see that discontent is mainly with economic conditions. The Conservative too often pins his faith to crude protectionist fallacies, especially dangerous in a country which depends on exports to pay for its food and raw materials. The fact that the electorate periodically rejects protection on quite wrong grounds does not prove it false in theory, but does make it impossible in practice.

The Socialist, while paying lip-service to free trade, is at heart a protectionist of an even more dangerous kind, and, in his practical embodiment of Trade Unions and the Labour Party, seems willing to protect any form of labour monopoly which can be established. The Labour Party must remain a "class" party, devoted to narrow class interests, and even to them in a short-sighted manner, so long as its constitution is based almost exclusively on the economic organization of the manual workers. If proof be needed, we have but to remember May 1926, when not one of the Parliamentary Labour leaders had the courage to denounce the general strike, which, had it succeeded, must have destroyed the basis of Parliamentary power.

The Socialist's main desire is to put the means of production under some form of collective control, with a view to a better distribution of its proceeds. He, too, overlooks facts which tell against him. The Socialist Governments which ruled various countries in Western Europe after the war, when faced with practical problems, could do nothing to realize their broader ideas. No one now advocates the old theory of State Socialism, and nothing but vague generalization has taken its place. It is curious that the British Labour Party has made its great advance in numbers at a time when the intellectual basis of Socialism, which theory of politics it has rashly adopted, has openly failed.

Except in limited spheres which expand only slowly with knowledge and experience, most available evidence shows that democratic collective management of almost any kind means a greater increase in direct costs of production than is balanced by more widely extended control and the elimination of private profit. For instance, the success of the democratic co-operative movement of consumers does but emphasize its repeated failure among associations of producers, owing to inevitable want of discipline and of other factors in efficient management. Therefore, to expand collective action in directions that need individual responsibility and prompt initiative, or to expand it in any direction faster than the wisdom and administrative capacity of Public Authorities is growing, leads to a net loss in efficiency, which would soon lower all round the standards of life which collective control had equalized.

Indeed, equality in the distribution of the national income would do much less than is commonly supposed.

Sir Josiah Stamp, the eminent economist and statistician, has proved by facts and figures which have never been controverted that, if the excess of all incomes above £250 a year in England were distributed equally among the families with incomes less than that amount, each family would obtain as a maximum an additional 5s. a week for the first year and less afterwards. An infinitely greater effect would be obtained by all pulling together to increase production and to distribute fairly the new wealth thus created.

It must not be forgotten that, in present conditions, every approach towards equality of income, if it be made at the expense of the saving classes, has drawbacks to set against its advantages. It leads to an increase in present happiness, since the need of those who gain is greater than that of those who lose, and leads also to an increased demand for consumable goods and thus to a growth in immediate trade, but it diminishes the amount saved, and therefore the provision for the accumulation of capital and increased employment in the future. It swells immediate wages at the expense of future wages, it is mortgaging the future for the sake of the present.

A gradual approach to more equal incomes is desirable and is going on; to force it too rapidly might be disastrous. On the other hand, a growth in national production, if well distributed, increases both immediate wages and the rate of accumulation of capital, both present prosperity and future demand for labour. With growing production, and an equivalent expansion in currency and credit, rising wages are beneficial to all, for they are needed to increase the circulation of currency in proportion to output, and prevent the depression which inevitably follows a deficiency of purchasing power. On the other hand, if production be restricted and be not increasing, a rise in wages leads to unemployment. Let me again quote from Sir Josiah Stamp[1]:

Since the war every rise and fall in the rate of real (as distinct from money) wages which we have attempted to pay have been correlated with statistical exactness to the fall and rise of employment, because both were over-ruled by the rise and fall of total production. If the total production was not changing, then the more we try to pay people above the economic rate, the fewer could remain in employment to get it.

Thus we return to the importance of improving the efficiency of our whole industrial machine, and removing the hindrances to its working. As already shown above, among the worst of these hindrances is instability in the value of money, only to be cured by international action. Secondly, partly as a consequence of recent deflation, are the high relative costs in the sheltered industries, which fall so heavily on the unsheltered. Thirdly, there is the appalling waste due to industrial disputes, and trade union and other restrictions which lower output, reduce possible wages and prevent the transfer of labour from depressed to flourishing trades. The organization which the manual workers have created has been useful to them in other ways, and, by making possible collective agreements, should be of national advantage, but it is by these restrictions doing much to prevent wage-earners benefiting by modern methods of improving the conditions of life. Trade unions should follow the American example and

[1] *The Times*, October 19th, 1926.

turn from barren strife to help in the acquisition of its capital by the workers in each industry and thus of industrial ownership and control.

But too often the managers of industry are also to blame. Their treatment of their men in the past, if not in the present, their frequent blindness to the national advantages of a policy of high wage-earnings, have been among the causes of trades-union restrictions. "Business men" are usually very efficient within certain limits, but some are curiously unable to grasp the import of broad economic changes, to meet which the reorganization of a whole industry, perhaps on co-operative lines, may be necessary. As knowledge, transport and communications improve, combination should and must gradually replace competition as the basic principle of industry.

Again, the Government and their advisers show frequent signs of that need for a "General Economic Staff," the formation of which has often been suggested. Decisions seem to be taken on questions of policy with no proper appreciation of the probable or certain economic results. For example, the return to the pre-war gold standard may possibly, on balance, have been to the national advantage, but it was undertaken with no provision for the inevitable consequences—trouble in the coal and other heavy export trades and an increase in agricultural depression.

Conditions can be improved slowly with the help of science, but most of our present discontents are really with the unalterable facts of existence. It is useless to kick against the pricks. Let us accept what is inevitable, and help each other to make the best of this sorry world.

It can be considerably bettered if we will but act on knowledge already won. Change and development are always
needed, on the land and everywhere else, but let us try
to analyse the causes of our troubles and the possible
conditions of improvement before we rush in with revolutionary plans where wiser men, at any rate, would
fear to tread without a previous dispassionate and ex-
haustive survey of the ground.

If we wish to go further and really help the farmer and
farm labourer, and the workers in other trades suffering
from low prices and high costs, we might usefully do
what in us lies to diminish the inequalities in charges
between the sheltered and unsheltered industries. To
this essential point we are constantly brought back.
Everyone is in favour of the highest possible wages.
Personally, I regard it as the chief end of economic
science to examine the conditions necessary to raise the
standard of life of the whole people, and the chief business of politics to secure them. But a man cannot get
a fair start if one of his legs is crippled, and British
industry cannot make up lost ground and begin to do
better while the unsheltered industries are carrying their
present load of costs. The rise in the value of the pound
during 1924–5 has cheapened external goods, and the
cost of living and of other things has fallen somewhat.
If internal goods and services can be cheapened too,
the cost of living would fall further. This might be done
by increasing the output per man by better organization or more piece-work in some trades. In others this
may not be possible, and the only cure is a temporary
fall in money wages. A slight lowering in nominal wages
in the sheltered industries would be partly recouped in

an increase in what these wages would buy; the fall in real wages would be much less.

In May 1926 railwaymen and other transport workers came out on strike in a futile attempt to support the coal-miners. The general strike could, of course, only do harm to the miners and everyone else, while, wittingly or unwittingly, it was a blow at the life and liberty of the nation which had to be parried at all costs. The real acid test of goodwill to the miners and to the low-paid wage-earners in engineering or agriculture is this—are the well-paid men in sheltered industries willing to accept temporarily somewhat smaller nominal incomes in order that costs of transport, distribution, etc., may be lowered, and thus the unsheltered industries, exposed to world competition, recover stability and become once more able to pay good wages? If so there is hope for us yet.

There is no fear of permanent burdens. The gold value of the pound has now finished its recovery, and only one more effort is needed to get the wheels of industry into gear again. When we have made that effort, the nation can start in confidence and goodwill to work together to apply the recent developments in science and in business organization, and make high wages and large output combine to give good profits and shorter hours of labour.

A drop of say 5 per cent. in sheltered wages (less in real earnings), with an equivalent lowering in other costs, especially an increase in output in such activities as building, might justify a rise of perhaps 10 per cent. in the present low unsheltered wages, and put agriculture and the export trades into a sound condition again. If employers and workmen then agreed with mutual goodwill to do all that the knowledge of each suggested to increase production, and remove all restrictions that stood in the way of an increase in output or mobility of labour, men could move into parts of the country where expansion is going on, the temporary loss in wages would rapidly be more than made up, and a general rise in the standard of life would be possible, a rise not confined to railwaymen and municipal servants, but extending also to the coal-miner who works for export, the skilled engineer, and the agricultural labourer.

CHAPTER XI

# THE OUTLOOK FOR AGRICULTURAL PRICES AND COSTS

Now that we know the causes of agricultural prosperity and adversity in the past, can we predict how they will work in the future, and foresee what will be the changes in prices and costs?

Let us first consider prices. The pound is once more linked with gold, and, unless there be another world convulsion, the link is not likely to be broken in the present state of financial opinion. But, even if the link is assumed to be of the same nature as before, the future relation between the supply of gold and the demand for it may be different. Some authorities expect that the economies in the use of the metal, effected by paper currencies and modern systems of credit, will enable supply to outrun demand, and tend to raise prices. But it seems that the higher present price levels compared with those of 1913 have already absorbed these economies and represent their effect. It is probable that the return of one country after another to a gold standard and the consequent call for reserves, even if some parts of these reserves are kept in foreign exchange values, will so increase the demand for gold that the supply will run short, as it did from 1874 to 1896, prices if uncontrolled will fall, and a long period of industrial and agricultural depression will follow[1].

[1] See, for example, the *Report of the Indian Currency Commission*, and an abstract of a Swedish article by Professor Cassel, given in *The Times*, 1 November 1926.

But we must not assume that the link which connects a currency with gold will always be the same. The volume of currency and credit which it is safe to erect on a given basis of gold varies as banking methods and commercial habits change. In this country the cost of bank credit to borrowers, as explained in Chapter v, is largely controlled by the rate of discount fixed from time to time by the Bank of England. In America such matters are chiefly managed by the Federal Reserve Board in New York, and the action of that Board since 1922 shows how new ideas can affect monetary policy.

In 1922 New York was the best market for gold; that is to say, gold could there be exchanged freely and directly for currency, and by means of that currency would indirectly buy more goods and services than it could secure elsewhere. Consequently gold poured into America. Treated in the old way, this gold would have been made the basis for a rapid expansion of credit. Prices would have risen, and there would have been an unprecedented boom in trade, followed someday by an inevitable slump. Deliberately, the Federal Reserve Board refrained from thus using their gold. By law they had to buy it when offered at a fixed dollar rate, but, to prevent a rise in the general price level, they buried it in their vaults as they bought it, and neither put it into circulation nor expanded credit in proportion. Since then, it is true, the gold has been used as the basis for an expansion of credit to meet the demands of a naturally growing industry, and prevent that growth being checked by the usual cause—a deficiency in purchasing power. The consequence is unexampled prosperity in the United States.

This example shows how, in a special case, even with a gold standard, currency and credit can be managed so as to keep prices approximately constant in face of an excess of gold. With a shortage of gold, it may not be so easy. To increase the ratio of currency and credit to reserves might be effective, but the limit of safety might soon be reached if one country acted alone. Now that other countries are also based on gold, international agreement may be possible, and even London and New York acting together could probably do much to stabilize the general level of world prices. The Genoa Conference of 1922 recommended international action to this end.

Whether the countries whose Governments agreed to that recommendation, or even England and America, will combine to maintain a stable price level remains to be seen. If they should do so, their efforts will probably be directed at first only towards smoothing out the wave-like fluctuations of good and bad trade which seem normally to recur every few years. The long-term drift of prices, due to variations in the volume and methods of business, and in the demand for and supply of gold, the drift which peculiarly affects agriculture, will probably remain, and, as we have seen, the direction of that drift in the immediate future appears likely to be downward, owing to a shortage of gold.

Now let us turn to the factors which affect agriculture specifically and not other industries. The separate causes of rise and fall of price in one kind of produce, the glut of potatoes due to a good season, or a scarcity of malting barley due to a bad summer, will always be liable to occur, and can perhaps be met best by co-operation between farmers for "orderly marketing," whereby the supply of produce is adjusted to the demands of the market.

"Orderly marketing" has succeeded in several foreign countries in trades which work chiefly for export. A brief account of present achievements in the United States, Canada and Australia, as well as in the better known instance of Denmark, will be found in Mr Enfield's work, *The Agricultural Crisis,* and in the *Report* of the Committee on the Stabilization of Agricultural Prices. A more critical description is given in Mr J. A. Venn's book on Agricultural Economics, together with a valuable discussion as to how far foreign and colonial methods are applicable to English conditions.

The main factors in success are stated by the Stabilization Committee to be the following:

(1) The organization must be on a "commodity basis," that is to say, a single co-operative organization must deal with only one commodity or with commodities closely allied to each other.

(2) The local co-operative associations must be federated to a central association in order to concentrate in a central authority the control of a sufficiently large volume of produce.

(3) Members must bind themselves under contract to supply the whole or a definite proportion of their produce prepared for market to the society of which they are members.

These principles seem first to have been put into complete operation in America by the California Fruit Growers' Exchange, and have since been widely adopted. In this way, American farmers have supplied the market steadily with certain kinds of produce, freed themselves from the disastrous effects of a glut and appropriated some of the profits of the middlemen.

But, as Mr Venn points out, an organization which suits American or Canadian growers, chiefly concerned with one crop, whether wheat or fruit, or Danes specializing in dairy produce for export, may not be equally effective for English farmers raising a variety of things for the home market, which, being within reach, is always tempting them to break away from their association. Nevertheless, Mr Venn considers that similar methods might succeed with poultry, eggs and butter, especially from smallholdings, and with the produce of market gardening areas, such as Cornwall, Bedfordshire or the Isle of Ely, or with that of fruit growers in the Vale of Evesham, in Kent or in Cambridgeshire. Beginning in this way, "orderly marketing" might be extended to other things as experience accumulated. Here, at all events, no Government organization is needed; farmers can, if they will, explore the possibilities for themselves, and financial help can be obtained from the fund provided to assist Empire marketing.

Much has already been done in one kind of agricultural produce, namely, milk. The price of milk is now fixed from time to time by negotiations between representatives of the National Farmers' Union and of the milk-distributing firms. The latter assume responsibility for disposing of surplus milk, and convert it into cheese or other products. Thus the farmer is assured of a market for his whole output, at a price known beforehand. He may not always get as much as he ought, indeed the very strength of the distributors' combination tends to beat down the price, but, on the other hand, the stability is a great gain.

There is another side to this problem of pressing importance, the solution of which is even more difficult. All enquiries into the subject have brought out clearly the great difference between the prices paid by the consumer and those received by the producer. The consideration of this question must some day be faced; information is available as to the facts, but few useful suggestions have been made for improvement.

Many difficulties stand in the way. It sometimes pays a retail shop better to sell a small quantity of, for example, vegetables at a high price rather than to handle more stuff at a lower figure. This may result in a fancy price in the shop, when the grower finds his produce almost unsaleable. Bakers and their workmen seem to get too big a share of the price of a loaf, and the milkman who delivers in a London street is much better paid than the highly skilled man who milks and tends the cows. Here we touch again the problem of the sheltered and unsheltered industries, which we have dealt with already in previous chapters. Enquiries into markets and marketing are being made by our agricultural economists. An interesting report by Mr F. J. Prewett has been published by the Oxford Institute for Research in Agricultural Economics. In this *Report*, criticism of existing methods and suggestions for improvement will be found. A co-operative market for livestock has been started at Banbury with promising results. Such activities can be and should be extended.

Bold proposals have been made from several sources for taking the whole problem of marketing out of the

hands of the farmer. Since prices are the chief factor
in agricultural prosperity, it is natural that those who
see its importance to the nation should, from time to
time, have wished to fix prices at a remunerative level
by legislative action. This idea is indeed only a return
to mediaeval views. A fair price and a living wage, both
enforced by statute, were parts of the general conception
of economic polity current in the Middle Ages, and only
passed away as the mediaeval social and economic
framework proved too rigid to allow the expansion of
industry which ushered in modern times. The doctrine
of *laissez-faire* undoubtedly gave the necessary freedom
for national wealth to grow in a time of transition.
Indeed it allowed industry to expand too fast for
administrative control, and a slower pace would have
been better in the broader interests of national welfare.
It has now ceased even to be orthodox. All parties
agree that some deliberate control of economic factors
is necessary, and only differ about its direction and
amount. The chief danger seems to be that legislative
control and effective trade union action are more and
more applied in ways which increase the costs of production, while much less is done, or perhaps can be done,
to cheapen them.

During the war, the inevitable rise in values forced the
Government to fix the price of foodstuffs. Quite rightly,
the consumer was thus saved at least £100,000,000 at
the expense of the farmer. It was supremely unlucky
that the attempt to repay this debt by insuring the
farmer against an unremunerative price of wheat was
inaugurated on the eve of the violent fall in agricultural
prices in 1921, and that a guaranteed price was fixed

which almost at once became impossibly high. Frightened by the cost to the nation, the Government hastily repealed the Corn Production Act. The experiment in stability thus failed. The failure was accidental, but it has immensely increased the difficulties of another attempt.

Nevertheless schemes new and old continue to be proposed. It is useless to suggest anything which will raise appreciably the cost of food. A predominantly urban population resolutely rejects at the polls any such policy. Protection is thus ruled out, while the heavy subsidies necessary to produce the same effect on agriculture would overburden an Exchequer already in deep water.

A scheme which professes to avoid the worst of these difficulties has been put forward from several sides[1]. It involves the national control of imports, at all events of wheat and meat. A Government department, or an independent non-profit-making Wheat Trust (on the model of the Wheat Commission during the war), would control all imports. It would survey home production, and obtain reports from the overseas dominions. If these sources of supply were inadequate, it would purchase enough foreign wheat to meet the probable consumption.

This scheme was pressed on the attention of the Imperial Economic Conference on October 9th, 1923, by the Rt Hon. S. M. Bruce, Prime Minister of the Commonwealth of Australia. In the debate on Imperial

[1] *Report of the Committee on Stabilization of Agricultural Prices*, p. 69 *et seq.*; *The Land and the Nation*, p. 437; *A Labour Policy on Agriculture*, p. 15 *et seq.*

Preference on June 18th, 1924, Mr Baldwin asked if some such course of action were not possible, and Mr Snowden, then Chancellor of the Exchequer in the Labour Government, replied sympathetically that the Government would examine the proposal. The scheme is also supported in *A National Rural Policy*, published for a Special Committee on Rural Reconstruction by the Labour Publishing Company, and somewhat similar suggestions have been made by the Independent Labour Party, in the Minority Report of the Food Commission, and in the official Labour pamphlet.

Indeed, as regards this question, the authors of the Labour pamphlet are far more alive to facts than are the writers of the Green Book and those who took part in the Liberal Conference. In a section headed "The Importance of Marketing" the Labour pamphlet says:

Until the War agricultural reformers concerned themselves chiefly with the problem of landownership and tenure. The monopoly of land, the burden of rents, and the control over agricultural development exercised by a semi-feudal class, constituted the most obvious hindrance to agricultural progress. But vital as is a change in the system of landownership, the last ten years have shown clearly enough that the question of marketing and prices is just as important. When agricultural land is nationalized, and a system of land administration brought into operation which will encourage and support improved methods of production and cultivation, the price received by the cultivator for his produce will still be by far the most important factor in determining whether the industry is prosperous or the reverse: and, in determining the prices actually received by the farmer, the methods of marketing and distribution are of primary importance. There is little dispute on the proposition that the marketing and distribution of the products of British soil fall far short of an efficient standard. The farmers cannot be held primarily responsible for this, for their chief concern should be with actual production. Between producer and consumer there has grown up an elaborate system of merchants and distributors which, as Commission after Commission has reported, absorbs far too large a percentage of the total paid by the consumer. The profits of speculators, of middlemen and of retailers, coupled with the currency and credit policies of Governments and banks, exercise a profound influence on the farmer's income and the labourer's wage.

Now the reference to "monopoly," "semi-feudal class" and so on, may be dismissed. A party which professes socialist opinions is in duty bound to make them, and we have dealt with such ideas faithfully in the preceding pages. But the recognition of marketing and price as the real outstanding agricultural problem shows the Labour pamphlet, short as it is, to be a more useful and practical contribution to the subject than the longer and more doctrinaire Liberal treatise. In the Green Book of 584 pages about 9 pages are devoted to marketing and the stabilization of prices, the subject being dismissed with a refusal to undertake measures of stabilization and a pious aspiration that "on the institution of cultivating tenure, the Government shall set up a Commission with wide powers to deal with the obstructions which prevent agriculture short-circuiting and simplifying its distributive services." On the other hand, in the Labour pamphlet of 40 pages no less that 11 (equal in number of words to 22 of the Green Book) are devoted to an analysis of the problem of price fluctuations and discrepancies, and a national scheme, resembling that suggested by Mr Bruce and the Committee on Stabilization, advocated.

Let us then examine these proposals in more detail. Besides controlling all imports of wheat, the Board or Trust, directly or indirectly, would fix prices for both British and foreign standard qualities of grain. And here different modifications of the scheme appear. A *National Rural Policy* proposes that the whole home crop also should be handled by a national distributive organization, and requires that the price of British wheat should be fixed, and fixed apparently with sole reference to providing an adequate profit to the farmer and an adequate wage for his men. Unless unexpected economies are effected by large-scale purchases of colonial and foreign corn, this involves some rise of price to the consumer as long as world prices are low. Since it only concerns the relatively small quantity of home-grown wheat, the rise would be less than under ordinary schemes of protection. But unless foreign grain rose in price proportionately, bakers would buy less English wheat. The object of the scheme could then only be secured by means of a subsidy.

On the other hand, the suggestions made by Mr Bruce, elaborated by the Ministry's Committee on the Stabilization of Agricultural Prices, and accepted in the Labour pamphlet, contemplate merely the avoidance of short-term fluctuations in price by controlling the imports of wheat and flour. If prices fell, foreign wheat would be held up. If they rose, foreign wheat would be released, and thus indirectly British prices would be controlled. The Committee on Stabilization say:

The object of the Board would be to make neither a profit nor a loss on its transactions as a whole, but to sell imported wheat and flour at steady prices which would as nearly as possible correspond with the average cost. Its buying policy might also be designed with the object of steadying prices in the world market, and, since the British demand constitutes a considerable fraction of the total demand of importing countries, it might possibly prove that the British Board as the largest single buyer in the market would exercise a determining influence on the world prices. As a further means of steadying prices, the Board might seek to enter into contracts with Dominion Governments and producers' pools for the bulk purchase of their crops for a year ahead. Provision might also be made in such contracts whereby any difference between the contract price and the average world price over a period should be apportioned between the two parties.

After the Board had been operating for a few months and had had time to make a rough estimate of the probable course of prices and supplies, it might be prepared to announce the basis of prices at which it would sell imported wheat in the home market for a definite period ahead. Only experience can determine for what periods it could safely keep the price steady; it is fair to assume, however, that the Board would be able to avoid constant changes of price and succeed in smoothing out the monthly and seasonal fluctuations that now take place. To put the case at its lowest, changes in wheat prices should not be more frequent than, say, changes in petrol prices under the organized marketing system of the large oil companies. Even a modest approach to greater stability in the price of wheat would confer a substantial benefit on British farmers.

It has sometimes been urged that the effect of concentrating in the hands of a single buying agency the total demand for imported wheat might be to accentuate fluctuations in world prices rather than to steady them. If the Board held off the market when prices were falling, it might aggravate the fall; and *vice versa* if it bought large quantities on a rising market (which it might naturally be tempted to do), the effect would be to exaggerate the rise. This view implies that the Board would endeavour to "beat the market" by choosing its own time to buy. But any scheme for centralized purchase on the lines we are here considering, presupposes that the Board's buying policy would be actuated by the same general desire for stability, which now actuates the selling policy (for example) of the central overseas selling agency for the Australian Co-operative Organizations. It would aim at buying steadily....

We think it not unreasonable to imagine that, as regards the British Empire, developments upon these lines might lead to a systematic interchange of information amongst the countries of the Empire in regard to requirements, stocks, purchases, sales, etc., with the object of pursuing a common Imperial policy and thus achieving what we understand to be the fundamental idea in Mr Bruce's proposal, namely, an orderly marketing system for Empire grain[1].

With this variety of the scheme it will be seen that no attempt is made directly to fix the price or to interfere in any way with the marketing of English wheat. Increased stability of home prices would however be obtained, for English wheat tends to a value a little below that of imported wheat and foreign supplies would be adjusted to secure stability. This indirect method is also favoured by the Minority Report of the Food Commission. Somewhat similar proposals are also made to regulate the imports and prices of meat.

I think there is more to be said for the national organization or at all events supervision of distribution than of production. The middleman is in too strong an economic position with reference both to the producer and the consumer, and perhaps needs some control. A national organization might certainly avoid some of the fluctuations in price from which at present both producer and consumer suffer. Such fluctuations undoubtedly involve a waste of wealth. But national management would probably involve loss in efficiency,

---

[1] *Loc. cit.* pp. 73, 74, 75.

and whether loss or gain would be the greater on balance, it is, I think, impossible to predict.

It is true that of the sums paid by the consumer too little reaches the producer, but the problem is not so simple as some socialist reformers think. For instance, Mr Philips Price considers that, if the State had bought out the United Dairies at par just after the war, "the whole of the subsequent profits would have either gone to the community or could have been used to cheapen the retail price of milk[1]."

Such an idea ignores the fact that profits, especially in a business that involves complex organization, depend on a very narrow margin between receipts and expenditure. Efficient and shrewd management and constant watchfulness are needed to turn a loss into a profit, and both loss and profit are cumulative. If the State had taken over the distribution of milk, producer and consumer might have gained, but it is very likely that the profits made by United Dairies would have vanished into thin air, and the State have made a loss.

I hold no brief for the middlemen, and agree that the farmer needs help against them. I think a better case can be made out for State action in distributing trades than in most others, but the case rests on the benefits of stability to producer and consumer, and not on the fact that some distributing concerns make profits of 12 or 15 per cent. on their capital. After all, a country can only increase in capital resources and national income, and improve the average standard of life of its people, if its industries are carried on at a profit and the excess reinvested—a doctrine old-fashioned but none

[1] *The Guardian*, January 1926.

the less true. Any business which makes good profits is a national asset.

The authors of *The Land and the Nation* declare against the policy altogether, on grounds of diminished national income and consequent unemployment, administrative difficulties, the danger of political pressure, international complications, and failure to benefit those who work on the land[1]. The arguments they adduce are strong, but in my opinion those that apply are not so strong when aimed at the control of foreign imports as they are when used against their own scheme of nationalizing the ownership of agricultural land. Reasons have already been given for believing that the Liberal scheme would injure, and not benefit, both agriculture and the national Exchequer, but any approach to stabilization of prices, even if on balance it involved a loss to the Exchequer, would at least help the agricultural industry.

My own criticism would be on different lines. I think the possibility of organizing markets should be explored carefully and thoroughly both by farmers and the Government. The difficulties are great; any interference with the play of natural economic movements is dangerous, though it is sometimes more dangerous to leave them alone. But the tendency is inevitably towards unification even in private enterprise. Where home growers supply the chief part of the market, the power is in their own hands. Where imports predominate, national action is needed for stabilization. Much evidence of its methods and results is now available, and a close analysis might show that its extension under national control to some kinds of imported agricultural

---

[1] *The Land and the Nation*, pp. 443, 444.

produce is possible. Of course the danger in initiating any such scheme is that one industry after another should be nationalized, and their profits, on which ultimately national solvency and welfare depend, destroyed. In a few industries it is possible that the gain in other directions might be worth the loss, but if all profits be turned into losses, complete national ruin will be the result. Perhaps the worst dangers of national trading might be avoided by working through non-profit-making trusts, subject to national supervision. With regard to the marketing of agricultural produce, at any rate an experiment could be tried without committing the country to continue it permanently should it prove a failure. Thus I am in favour of (1) expert enquiry, and, if that proves favourable, (2) a tentative experiment.

Too much benefit, however, must not be expected. The organization of markets, to which the name of stabilization of prices is often given, is not complete stabilization at all. It leaves untouched the danger of a long-term fall in prices which, as we have seen, unless there should arise a world-shortage of food, may set in from monetary causes—in present conditions an increased demand for, or reduced supply of, gold. When gold is plentiful and prices rise, agriculture flourishes, and, when gold is scarce and prices fall, agriculture inevitably decays. It is this slow change, and not the fluctuations of markets, which causes periodic and uncontrollable general depression. If our standards of weight or length depended on the amount of (say) copper which had been mined, both science and industry would suffer. Industry suffers in a similar way from variations in our standard of value. The dangers of departing from the gold standard may at present be too great for bankers to contemplate. But nothing is gained by closing our eyes to the fact that to let the standard of value be ruled by one arbitrary commodity is Relativity in the wrong place. We are putting ourselves at the mercy of chance changes in the world's output of gold, or of an increase in the demand from India for purposes of hoarding. The probable lives of most South African mines are said to be only about twenty years; unless a new gold field is discovered, which is unlikely, an acute shortage of gold will then develope. We must some day find a more stable and scientific standard than gold, based probably on an index number of the most important commodities[1]. Till that day comes, or till the world's output of food begins to fall short, the organization of marketing may smooth away some of the farmer's minor troubles, but it will not banish periodic general depressions in agriculture.

More modest proposals for the alleviation of our present distresses have often been made. One of them involved a duty on malting barley. This was accepted at first in principle by members of the Government of the day, but was found to be impracticable under existing circumstances owing to unforeseen legislative difficulties.

An interesting suggestion is that a maximum price, adjusted from time to time, should be put on ordinary bread. Since English wheat is cheaper than foreign, this

[1] See, for example, *A Tract on Monetary Reform*, by J. Maynard Keynes, London, 1923.

is expected to increase the proportion of English wheat used by bakers for the normal loaf, while leaving them free to use and charge what they like for fancy bread[1].

But all these ideas depend for their adoption on outside action, political or other, and farmers will perhaps do well not to rely on seeing them put into force. Let us therefore return to probabilities of agricultural revival under the unrestricted play of economic forces.

In his address to the Agricultural and Economic Sections of the British Association in 1926 Sir Daniel Hall gave reasons to believe that there was not enough new agricultural land available to meet the needs of a still expanding population at present prices. This conclusion is not universally accepted, but he also referred to another factor which will affect agricultural prices generally—the increasing demand of agricultural workers of all grades and all countries for higher remuneration, and the consequent rise in the cost of growing food. England is still predominantly a land of tenant farmers and of larger holdings than are usual elsewhere, at all events in Europe. Large farms are for the most part worked by hired labour, and agricultural wages, low though they are compared with those in our own sheltered industries, are higher than in any other European country. But England is exceptional. Elsewhere farming is mostly carried on by the owners of the soil, and, in old countries, chiefly in what we should call smallholdings, by men and their families who work harder and for longer hours than do English labourers.

---

[1] Personally, as I much prefer, and cannot obtain, bread made chiefly from English wheat, and dislike the puffy white bread now so popular, I welcome this proposal.

Peasant husbandry is still responsible for a large part of the agricultural output of the world. In spite of all we hear of the benefits of smallholdings, the standard of life of these peasant husbandmen is low, in most parts of the world lower than that of our agricultural labourers, and it depresses the standard of those with whom they compete.

But, as the wealth of the world increases, and a knowledge of the amenities of civilized life extends, there is a steady movement to the towns. The fraction of the population engaged in agriculture steadily diminishes, not only here, but in all other lands, and since the war this tendency has spread to peasant proprietors. It seems likely that the total world output of food will therefore become less compared with requirements, and its price, relatively to that of commodities made in factories, will rise, until a standard of life comparable with that of the urban workman becomes possible for the peasant husbandman. This is another reason why Sir Daniel Hall expects the price of agricultural produce generally to rise in the near future, and thinks that the day of cheap food is almost over.

In summing up the results of our survey of the prospects of agricultural prices during the next few years, we have to consider both causes which affect the general price level of all commodities, and also such tendencies as those pointed out by Sir Daniel Hall, which may affect agricultural prices relatively to others. It is dangerous to prophecy, especially in matters economic, which involve the uncertain element of human psychology. All knowledge, even that of science, is but an affair of probability, and logically we can only express the intensity of our beliefs in terms of a bet. If all Governments were ruled by reason, doubtless the general price level would be stabilized by international action, but, in our present world, it would perhaps be wise to put the chances as at least 3 to 1 against this action being taken. Twice in the last century prices were raised and agriculture revived by the discovery of new goldfields. It would be rash to hope that any more such large discoveries will be made in the near future. If general prices are not stabilized deliberately, they will almost certainly slowly fall.

Now suppose we guess that the chances are 2 to 1 in favour of Sir Daniel Hall being right; and assume that the causes he describes may produce about as much effect as a probable change in the balance of gold supply and demand. We can then calculate the relative probability of agricultural prices rising as 2, of their remaining about constant as 7, and of their falling as 3. On the assumptions set forth above, it looks as though we might fairly bet 7 to 5 that agricultural prices will remain about steady during the next few years, while the odds seem about 5 to 1 against a considerable rise and 3 to 1 against a heavy fall. But of course the accuracy of this estimate depends on the validity of the assumptions on which it is based. No exact measure is possible; but to put down figures forces one to think out the real factors in a problem, and may perhaps give some guide to the order of the various probabilities involved. At all events, to estimate chances is safer than to make the glad confident statements of the professional politician or the amateur prophet.

Having now dealt with prices, we turn to the other factor in financial success or failure, namely, costs of production and distribution. How far can present costs be diminished, and how far will a diminution react in favour of the British farmer in competition with others?

A hundred years ago, after the Napoleonic Wars, costs were lowered chiefly at the expense of the rate of real wages of the labourer. Let us rejoice that this cruel and disastrous policy has not been revived. Real wages are now higher than in 1914, and the agricultural share of the burden of the Germanic War has fallen principally on the landowner who is better able to bear it. But labour costs are a higher fraction of total expenditure for the arable than for the grass land farmer, and hence to maintain wages in the face of falling prices favours the conversion of plough land to pasture. It also makes all other methods of decreasing costs more important.

In a study of agricultural economics, the logical place for the consideration of the whole great subject of agricultural research and education is under this heading of costs. If science can teach us how to make two blades of grass grow where one grew before, the economic effect is to reduce the cost of each blade, and of that weight of milk or meat derived from it.

The improvement in yields brought about by the introduction of root-crops, by enclosure, by selective breeding of animals, by artificial manures, is known to all students of history. Can we look for such gigantic steps in the future? Possibly not; yet great advance is taking place.

Plant breeding has been set on a scientific footing. Mendel's forgotten work on heredity, rediscovered by Bateson, has been applied to wheat by Sir Rowland Biffen, to barley by Dr Beaven, and bids fair to increase both yield and quality in all farm crops. Sir William Somerville has shown us how much poor grass land can be improved by basic slag and other phosphates. The use of milk records as a guide in the breeding of dairy cattle is increasing enormously the average output of our herds, and a lively controversy on the best food for milking-cows is developing light as well as heat. Professor T. B. Wood and his colleagues, by studies in animal nutrition, are cheapening the fattening of stock, and experiments carried out for the Royal Agricultural Society by Dr Thornton at Rothamsted are demonstrating that a crop like lucerne can be grown on land where it has failed before by inoculating the soil with appropriate bacteria. The diseases of plants and animals are being studied by scientific methods and in some cases got under control.

Efforts are being made to produce in England crops and their derivatives hitherto imported from abroad. The manufacture of beet-sugar is being subsidized, and the Ministry of Agriculture has carried out experiments on improvements in the method of manufacture. Sugar-beet is a useful crop for the arable farmer. Dr Harding, working for the Royal Agricultural Society and the Ministry, has developed a new process for extracting milk-sugar from cheese-whey, hitherto almost a waste product, and the application of this work in practice will help the dairy industry.

Much research being done is still in the stage of pure

science. From some of it practical developments are inevitable, but it is fatal to confine our work to points likely to be of immediate practical use. The greatest discoveries are made by men with a passion for knowledge for its own sake, with no thought of economic benefits. Faraday never dreamed of electrical engineering, nor Mendel of improved strains of wheat. Let us search for knowledge, and, in due time, all other things will be added unto us.

As an example of such work, let us take the studies of soil carried on at Rothamsted and elsewhere. Soil used to be regarded as a mixture of mineral constituents with decaying dead organic matter. We now know it to be a highly complex organic structure, living rather than dead. The mineral constituents of fertile soil are coated with jelly-like colloids, which absorb other bodies in the complicated reactions of surface chemistry. The physical and chemical nature of the soil depends on these reactions. Clay, for instance, is coagulated into larger particles by elements like calcium, and thus made more porous. If calcium be replaced by sodium, the clay becomes infertile, plastic and difficult to cultivate, and experiments show that more work is wasted in dragging ploughs and other implements through it.

Bacteria and other living organisms form an essential constituent of all soil, its properties depending on the number and species of its bacterial population, which varies from day to day, indeed from hour to hour. Counting and identification of such bodies is part of the ordinary routine of a soil laboratory. Some bacteria, found in the root-nodules of leguminous plants, convert the nitrogen of the air into nitrates, and so increase the fertility of the soil. Others perform different functions, which, whether known to be useful or not, must be studied if a complete knowledge of soil is to be obtained, and all the possibilities of that knowledge explored.

While great progress has been made in the application of chemistry to agriculture, the implements and methods of cultivation change but slowly. The plough is much as it was centuries ago, and the process of getting a tilth fit for a seed-bed remains the same, a nervous and uncertain coquetting with soil and weather. Only of quite recent years has a beginning been made in measuring accurately the power needed to draw different forms of plough and other cultivators, and to formulate the problems of tillage which can be attacked in the future by the agricultural engineer. Daring innovators and inventors are beginning to ask whether our long process of cultivation, which too often forms a pan of hard and impervious soil beneath the surface, cannot be shortened, whether indeed a prehistoric implement like the plough is really the last word, and whether the whole process of ploughing, cultivating and harrowing might not be carried through in one operation by some form of rotary tiller. The new Oxford Institute of Agricultural Engineering has a worthy field for its researches. Improvement in methods of cultivation may do much to diminish the necessary costs of the arable farmer. But let me once more point out that the ideal thus rightly aimed at is to diminish labour charges by developing the use of machinery driven by highly skilled and highly paid men. Probably fewer hands will be employed, but their economic status will be raised. Agricultural employment may be lessened, but national wealth, and therefore total employment, will be increased.

Scientific methods, formerly confined to the improvement of agriculture as a craft, are now being extended to its study as a business. Enquiries into markets and marketing methods have been dealt with under the head of prices, but agricultural economists are now investigating costs of farm operations as a whole and also costs of production of individual crops and animals. Only by proper costing accounts can farmers tell what part of their business is losing money, and what part is paying, and therefore worth developing. Only by a knowledge of the financial results reached by the leaders of agricultural practice, can ordinary farmers judge of the comparative success of their own efforts and how to improve them. Hence the importance of specimens of actual farm accounts, such as are issued periodically by Mr Venn from the Economics Branch of the Cambridge School of Agriculture, or the figures giving the costs of production of milk as investigated at the Agricultural College at Wye.

Agricultural research has no effect on production unless its results are worked out in practical conditions and made known to the farmer. This latter operation is now carried out in several ways. Through scientific articles and instruction at Universities and Colleges, information about new results reaches the County Agricultural Organizers, who are able to pass it on by personal intercourse with farmers. The sons of many landowners and a few large farmers take a scientific agricultural education, and come home to put it into practice. The farm newspapers are devoting more atten-

tion to agricultural research, and the new yearly volume
to be published by the Royal Agricultural Society should
be useful.

Agricultural practice should also be improved and
costs lowered as our general scheme of national agricultural education produces its full effect. Besides the long
courses in Universities and Colleges, taken in the year
1924–25 by a total of 1452 students, good work is also
done at the many Farm Institutes, and short winter
courses as well as County Council lectures and papers
at Farmers' Clubs, etc., are available for practical
farmers. All Committees of Enquiry agree that our
educational machinery is now excellent, and only needs
developing on existing lines. Facilities are available, but
there are not enough teachers, students or money.

On the wide question of education as part of the basis
for the reintegration of village life we shall speak in
the next chapter.

Whatever be the relation of rent to costs of production in abstract economic theory, the rent of English
land is wholly or mainly interest on the cost of
equipment (see page 122), and rent is certainly one
of the farmer's annual outgoings. But rent is already
too low for the maintenance and improvement of the
existing equipment of the land, and real rents cannot
be further reduced without danger. A reform of our
rating system is long overdue, but, even at present,
rates form a small proportion, perhaps 2 or 3 per cent.,
of a farmer's outgoings, so that much less relief can be
obtained in this direction than is often imagined.

Railway charges, though their increase since 1914 is
no more than the change in the purchasing power of

money, were and are a heavy burden, for they bear a high ratio to the value of most kinds of agricultural produce, especially of English produce, usually consigned in small quantities. Reform is needed, though it is hard to see how the railways can make much reduction as long as their own costs, especially for labour and coal, remain so high. The same remarks apply not only to the distribution of agricultural products, a subject with which we have already dealt, but also to that of farm supplies. Both in prices and costs, the farmer suffers heavily by the high charges of the sheltered industries.

### CHAPTER XII

## RURAL DEVELOPMENT

ALTHOUGH I have criticized the ambitious schemes of some would-be land reformers, I would not be thought to hold that all is for the best in the best of all possible countrysides. Landlords are impoverished; some have sold; others are either reducing too much their expenditure on repairs, or, as an alternative, have had to let their houses and are unable for the time to live on their estates and look after the people that dwell thereon. Arable farmers for three years lost on their trading accounts, and all farmers have had to write down their valuations. Labourers' wages, though higher than in other European countries, are still lower than everyone would wish; Wages Boards have in places caused a loss in allowances as against the gain in cash; cottages are often bad; too few men have possession of land.

Now, of course, all this is mainly the result of agricultural depression. Things will largely right themselves if and when there is a rise in prices or a fall in costs, including those in sheltered industries which at present bleed the farmer, and through him the landowner and labourer.

The most important step in all agricultural policy is for the Government to guard against increased depression by securing international action to stabilize the general price level as agreed at the Genoa Conference in 1922. Unless this is done, the growing demand for gold will cause a fall in prices—slow at first, but faster as the South African mines are worked out. Such an accelerating fall, due, not to a real and beneficial cheapening of production, but to a deficiency in purchasing power, must be disastrous to industry generally, and especially to agriculture.

This is the chief point in agricultural economics. But the whole rural problem involves other economic and social factors which can only be understood in the light of history.

How then are we to reach our object—the enrichment of the social and economic life of the village, and the replacement of the communal activities of the mediaeval manor, gone themselves beyond recall, with some modern substitute? Besides the organization and financial support, given through the Ministry, the Universities and the Colleges, to agricultural research and education, much is already being done—on one side by the equipment of County Council smallholdings to improve economic opportunities, and on the other by the establishment of Village Halls, Women's Institutes and Community Councils to develope social conscious-ness. Among other possibilities I think the follow-ing are important. Firstly, a further increase in the number of holdings, graduated in size, and credit facilities for those qualified men who might otherwise be unable to work them; secondly, the organization of large farms on a profit-sharing basis, as some few have been in the past; thirdly, the mutual support and intercourse that agricultural co-operation in its many possible forms may give; fourthly, when possible, the establishment of industries other than agricultural in the country; and, fifthly, the development of the village school, and perhaps in the future the village College, into a real centre of country life.

In the establishment of new industries, the development of electric supply should help. The present use of mechanical power on farms is not enough to encourage supply companies to run a network of high tension mains over purely agricultural areas. But, where there is a chain of villages from town to town, public supply becomes possible. Also in several places village power-stations have been erected and have proved a financial success, and, failing them, private installations are often possible. However obtained, good light and convenient power will add to the amenities of country life, and aid in the development of rural industries[1]. It should not be overlooked that such development may make an electric supply remunerative in places where the present demand is inadequate.

With regard to increasing the number of holdings, I have given reasons in Chapter IV for doubting whether the multiplication of small freeholds is possible or desirable. To this policy some Conservatives look for the complete solution of rural problems. I wish they were right, but I am almost sure they will be disappointed. Without some such restrictions as those of feudal times, small freeholds tend to disappear again. Lord Bledisloe holds that the success of Danish agriculture is partly due to the "magic of ownership," but Sir Henry Rew points to the evidence available in America, which shows that, in influencing production, other factors are much more important than systems of land tenure. The British

[1] See *Journal of the Royal Agricultural Society*, vol. LXXXV, 1924. British Association, 1926; Sir John Snell's address to Section G.

farmer at all events does not want to buy land[1]; he can use his capital more profitably by stocking it, and he likes to retain the power of moving to a larger or a smaller farm. What he really likes is to be tenant on a large estate, and, as the owner of a very small one, I quite see his point. I agree with Mr Lloyd George that the model system is "not ownership but tenancy on a good sound estate[2]."

The tenant farmer, as regards his landlord, can well look after himself. But to improve the present lot and future prospects of the labourer should be a real object of our efforts. Of course, here again, there is much exaggeration. On good farms he is not so badly off as some folks like to think, and, on the whole, is in a considerably better position than before the war. But his wages are still too low compared with those of other equally skilled men; it is often difficult for him to obtain land if he wants it; in the South of England there is too seldom a graduated ladder of holdings up which he can climb, and too few large farms which offer adequate positions as foremen and managers to good labourers.

To say that the average size of an English holding is 62 acres conveys little useful information, but the following tables[3] show clearly the distribution of the land among farms of different sizes.

It will be seen how large the number of smallholdings is. In spite of attempts to increase them, holdings below 20 acres are diminishing in number, but the "family farms" of from 20 to 150 acres are increasing. They fulfil an economic need.

---

[1] See p. 61, above.          [2] See p. 126, above.
[3] Venn's *Agricultural Economics*, pp. 61 and 68.

*Number of holdings in England and Wales*

| Size group | 1885 | 1895 | 1913 | 1921 |
|---|---|---|---|---|
| 1–5    acres | 114,273 | 97,818 | 92,302 | 81,217 |
| 5–20    ,, | 126,674 | 126,714 | 122,117 | 116,159 |
| 20–50    ,, | 73,472 | 74,846 | 78,027 | 80,967 |
| 50–100    ,, | 54,937 | 56,791 | 59,287 | 61,001 |
| 100–300    ,, | 67,024 | 68,277 | 69,431 | 67,842 |
| Above 300 ,, | 16,608 | 16,021 | 14,513 | 12,947 |
| Total | 452,988 | 440,467 | 435,677 | 420,133 |

| Size group | Total acreage 1913 | Total acreage 1921 |
|---|---|---|
| 1–5    acres | 285,000 | 253,000 |
| 5–20    ,, | 1,373,000 | 1,310,000 |
| 20–50    ,, | 2,623,000 | 2,720,000 |
| 50–100    ,, | 4,325,000 | 4,443,000 |
| 100–150    ,, | 3,942,000 | 3,955,000 |
| 150–300    ,, | 7,844,000 | 7,475,000 |
| Over 300    ,, | 6,737,000 | 5,988,000 |
| Total | 27,129,000 | 26,144,000 |

Where soil and markets are favourable, smallholdings should be multiplied, both by County Councils and by private landowners, for those exceptional men who have the will and the ability to use them. First there is the cottage with a garden. Even with the subsidy available for private enterprise, it is now impossible to build and let rural cottages on an economic footing. Before the war this was just possible in favourable conditions, and on one small estate in Devonshire it had been done in connection with a scheme for letting as many cottages as possible direct to the occupier[1]. Tied cottages are a necessary evil, and should be reduced to the lowest possible numbers. As costs come down, or more encouragement is given to the private builder, landowners may again be able to erect cottages. Meanwhile, they can give land to local authorities or sell it at low prices where cottages are needed. To those who refuse, compulsory powers may well be applied.

Next to the cottage and garden comes what may be called the cottage holding, in which personally I have much faith. It varies in size according to the soil, from half an acre to perhaps five acres, and helps to support a jobbing labourer who works for neighbouring farmers or others when they need him, and fills in time on his own land. This style of holding is being created in considerable numbers by the Forestry Commission, for work in woodlands in winter gives an ideal part-time occupation for cottage holders. The Government Bill, now before Parliament, also provides facilities for new cottage holdings.

Then we have the present smallholding, on which a hard worker can just make a living, though a sensible man supplements it with carting or other activities. The restriction of a statutory "smallholding" to one of 50 acres should be removed. Economic enquiries are bringing out the fact that the small farm which gives the largest net return per acre runs from 50 to 150 acres, according to soil and type of cultivation[2]. This is usually the economic unit for one pair of horses, and represents

---

[1] "Agricultural Labour and Rural Housing," W. C. D. and C. D. Whetham, *Edinburgh Review*, 1913, vol. ccxviii, p. 42.

[2] See p. 44, and Tables on p. 195.

the "family farm," which largely avoids labour troubles. It is a significant fact that holdings of this size are increasing in number at the expense of both smaller and larger ones.

So long as costs remain relatively high, as they are at present, landowners should be given subsidies towards building, as are Local Authorities. The high cost of building is largely due to the action of successive Governments—to their cowardice in dealing with the trade unions of building operatives, and the high subsidies they have had to give as a consequence towards the building of essential houses. These considerations, added to those about the fall in real rents, given in Chapter VII, make the claim of the country landowner very strong.

With adequate help towards the cost of houses and farm buildings, even on private estates, smallholdings can be formed where soil and markets give them a fair chance. The opposition to smallholdings comes generally not from landowners but from large farmers. Naturally no tenant likes to have land taken away for other purposes, and some smallholders are very troublesome to their neighbours in letting their stock stray in search of free meals. They are often a nuisance in this and other ways, but still must be encouraged for reasons of public policy.

County Councils already own much land—indeed in some counties, *e.g.* Cambridgeshire, the County Council is the largest landowner. Some of their smallholdings, equipped at the time of highest prices, have proved a heavy financial burden, but others are a success, and there is no reason why this form of tenure should not be extended and all restrictions about the size of hold-ings removed. It has the advantage of lending itself to the development of colonies of smallholdings, where adequate supervision and help and advice can most easily be arranged. When smallholders will consent to co-operate, this may become important. At present, for the most part, they are as individualist as large farmers, and there is more hope for them when scattered on private estates, so that each man can find a retail market in his own neighbourhood.

Many landowners are willing to finance promising men to start them in smallholdings, or guarantee a loan for them from the bank. I have done so myself and never lost thereby. An extended system of public credit will probably be arranged for applicants for County Council smallholdings, but it may be difficult for the Councils to select and supervise or grant public credit to men who wish to set up on private estates.

There has been much discussion on the subject of agricultural credit. A useful summary and suggestions for action will be found in a *Report* by Mr R. R. Enfield, which was published early in 1926 by the Ministry, as a basis for further consideration.

As in other departments of agricultural economics, some exaggeration of existing deficiencies will be found in the writings and speeches of ardent reformers, who wish to copy foreign methods without a clear under-standing either of the present facilities for borrowing, or the special requirements of the British farmer.

Where land is let by the owner to the cultivator, the need for long-term credit is met. The farmer can invest all his resources in live and dead moveable stock and working capital, because the landlord provides the land, buildings, and other fixed equipment. Thus one difficulty of the foreign owner husbandman does not afflict the English tenant farmer. It seems to be generally overlooked that the system of land tenure predominant in England solves in a most natural and economic way the problem of long-term credit so much discussed. In effect, the landowner lends to his tenant a sum of money equal to the capital value of the holding at 3 or 4 per cent. interest. What more could the newest and most democratic scheme hope to do for the farmer?

Nevertheless, the recent growth in occupying ownership creates a further need for long-term credit. Mr Enfield points out that, whereas urban industry has solved the problem by capitalizing future earning power and issuing debentures and shares of limited liability in Joint-Stock Companies, an owner farmer has to rely on private credit. Private mortgages are not always easy to negotiate, and involve a hampering uncertainty as to dates of repayment. The Agricultural Credits Act of 1923 has proved unsatisfactory owing to its restrictive clauses, and it is therefore proposed to establish a Central Land Bank. This bank would advance money on mortgage to farming owners through the introduction of the Joint-Stock Banks, with branches everywhere in close touch with farmers. On the general security of its total assets, the Land Bank would issue bonds to the public, which would thus be able to invest in agriculture with the same ease as in industry, or in Government and Local Loans. The mortgages would be repaid by instalments, at dates to suit the borrowers, and, as long as interest was paid, could not be called in.

The question of short-term credit, needed to finance the ordinary yearly business of the farm, is quite different. The large and progressive farmer already can and does obtain all the seasonal accommodation he requires from the Joint-Stock Banks. The common complaint that these banks are less sympathetic than the private banks they absorbed seems quite unfounded. The total amount of loans and overdrafts to farmers now outstanding is larger than in the days of the private banks. Whatever mistakes may be made and whatever be our criticism on certain broad issues of policy, anyone in close touch with modern banking must realize that in detail the Banks serve their customers well, and, with special reference to agriculture, provide adequately for the needs of the large farmer, who can offer reasonable personal or collateral security.

But the smaller farmers have not yet learnt how to borrow wisely, and often have no security to offer beyond their stock and growing crops. Sometimes a seasonal financial pressure forces them to sell produce at unfavourable times and leads to a glut on the market. Generally they resort to indirect methods of borrowing money. Firstly, rents due at Michaelmas and Lady Day are usually only paid in the middle of November and May. Thus every farmer, large or small, borrows £1 or £1. 10s. an acre for twice six weeks each year from his landlord free of interest, and, in times of difficulty, gets this time extended. Then the small man buys supplies on credit, sometimes long credit, from his dealer, who naturally charges higher prices to recoup himself, and often makes an open or covert stipulation that the produce when ready for market shall be sold to him. In this way, some farmers get into the power of dealers—occasionally with disastrous results. And, on the other hand, some dealers who wish to keep clear of these entanglements, find them necessary in competing with other dealers for farmers' custom. The system is bad for both sides.

Mr Enfield rightly urges that farmers should borrow openly from their banks at known cost, instead of indirectly from dealers at unknown and probably excessive charges. But he acknowledges that the difficulty is chiefly on the farmers' side. Many farmers appear unable to see that they are really borrowing money and paying interest to their dealers when buying on credit at higher than cash prices. As stated above, an educational campaign is needed.

Moreover, the publicity of the English bill of sale deters farmers from pledging their stock and crops. The legalization of the American form of chattel mortgage is therefore recommended, whereby, subject to the prior lien of rent, rates and taxes, a farmer could offer his stock and crops as security, the transaction being registered in records open to inspection by banks but not by the public. With this extension of present facilities, the Joint-Stock Banks could meet all legitimate requirements.

The official White Paper says[1]:

Credit is needed for the development of occupying ownership to which the Government attach great importance, and many farmers are seriously short of working capital, particularly if they have bought their farms. In these circumstances the Government are giving special consideration to the whole subject, and discussions are proceeding with a view to the preparation of a scheme on a sound commercial basis for

[1] *Agricultural Policy*, 1926, Cmd. 2581.

short-term credit, credit for improvements and credit for land purchase, with the object of bringing the general credit machinery more into line with the existing economic needs of the industry.

Till the result of the Government's deliberations is known, it is perhaps premature further to discuss the problem of credit. But it is worth while once more to point out that the much-abused landowner, as shown above, is already doing something towards its solution.

Profit-sharing schemes have been rather in abeyance since there have been no profits to share, but, when more normal times return, farming landowners are the most likely agriculturalists to develop a system in which some of them were pioneers. Corn growing is best done on the large scale, and, in some parts of the country, especially in those unsuitable for smallholdings, posts of increasing responsibility on large farms with a share of the profits should offer to competent men an economic ladder alternative to a graduated scale of holdings.

But, when all is said and done, the solution of our present agricultural difficulties is mainly an affair of prices and costs, and until Governments acknowledge this fact, and take steps to stabilize the general price level, they are but playing with the subject. The farmer, all unconscious of the underlying monetary causes, blames free trade for ills which present themselves to him in the guise of foreign competition. He has more to gain by stabilization of the general price level than most men. This he can do nothing to secure. Again he needs some assistance in his rather ineffectual struggles with dealers and middlemen. I have much sympathy with the idea that the farmer should be primarily an agriculturalist, and should somehow be relieved or relieve himself of the specialized work of marketing his produce. The large national scheme of controlling markets described above may not prove feasible, and it is difficult to get the farmer to co-operate for "orderly marketing" of the American or Danish type, but here, too, landowners have been useful in the past, and, if appealed to and encouraged, might be useful in the future in helping to carry out any scheme that may be adopted.

As an alternative to private ownership, especially in regions where resident landowners are few, I think a considerable extension of County Council ownership is desirable. County Councils might be empowered to buy land offered them for cash or land stock, and all restrictions about the size of holdings removed. Experience of public ownership would thus be increased, and its costs, advantages and drawbacks realized. The experimental method is always sound.

But the future of rural England does not depend on economic development only. Man does not live by bread alone. In spite of the poverty and privations of mediaeval life—far worse than our poorest now know—the mediaeval manor was a real community, with a corporate existence and a common outlook. It was a social as well as an economic unit.

The breakdown of the manor involved the disintegration of village society, and a real solution of the rural problem requires its reintegration in modern form. History and present conditions prevent its complete re-establishment on a religious basis, and a secular framework must be sought. In a remarkable speech to the British Association in 1926, Mr Duncan, representative of one of the Scottish workers' unions, pointed out that the successful rural life in Scandinavia was acknowledged to be largely due to adult education in the high schools, which brought young people together as a community during the winter sessions. Mr Duncan pleaded for a similar development in the village schools of this country. Education hitherto has been too individualistic, and the main defect in rural life is a disbelief in and disinclination for associated effort. Everywhere village schools, and in larger centres village colleges like those planned for Cambridgeshire by Mr Morris, might set themselves to educate children, and indeed men and women, in a corporate outlook, and teach them the great truth that we are indeed members one of another.

To anyone who knows well our present village life, Mr Duncan's criticism rings true. The great difficulty is to get the people to work together for any object. Local dissensions are so keen, neighbourly ill-feeling so universally enjoyed, that while villagers will usually follow the squire or the parson if he be willing to take a lead, they will not organize themselves for any form of corporate action.

Here we have another reason against nationalization of the land. Till a corporate spirit has been re-established, the natural leadership of a resident landowner is the best substitute for it, and the best chance of developing the spirit we hope for. Community Councils, Women's Institutes, village halls and clubs, the outward and visible signs of that inward and spiritual grace, grow more quickly and healthily where there is a resident landowner who can lend a helping hand.

But all these really practical measures can be carried through within the framework of the present system of tenure, more easily and effectively indeed than under the costly and inefficient Liberal and Labour schemes. The great and cardinal error of our would-be land reformers is that they have not used the landowner instead of abusing him. There are bad landlords as there are bad farmers, but there are also many progressive and intelligent landowners with a high standard of public duty. Modify the Acts which, by increasing unduly the farmer's security of tenure, have, as the Liberal Green Book allows, injured agriculture; give the landowner official support in dealing with bad farmers, some encouragement to initiate changes when they are needed, and peace from threats of confiscation. A well-considered scheme of rural development can then be carried through by his help, more easily, swiftly and successfully than without him. Landowners, already on the spot with local knowledge and a more detached outlook than ordinary farmers, are the best agents through whom to work. I believe more might be done to improve rural life by joint consultation and action between the Ministry of Agriculture and the Central Landowners' Association, working locally through landowners, than by all the democratic County Agricultural Authorities ever imagined by Liberal Land Leaguers or the Independent Labour Party.

But, with two of our three political parties now committed to the principle of land-nationalization, it would be foolish to ignore the possibility of a coalition between them to carry through some definite scheme. I think the loss, both social and economic, would exceed the gain. If, however, after due consideration, the nation should decide to take possession of the land, present landowners must help to make the new system as good as may be. Three conditions seem to me to be essential: the first, adequate compensation on equitable terms for the owners; the second, administration under the County Councils and the Ministry of Agriculture by competent professional land agents, as in the scheme of Orwin and Peel; and the third, a sensible, straightforward system of tenancy, either on yearly agreements or leases for definite periods, whereby the nation, which will have to bear much of the loss in bad years, may recover some of it when times are good. With the ever-widening residential areas round towns, I think the State, by taking over the whole of the land and thus establishing for the first time a real monopoly, might make complete expropriation pay its way. But, if so, unless there be a very great rise in agricultural prices, I feel sure that the profit will be made on building and other urban values, and not on national dealings with agricultural land.

Whatever the future may bring to rural England, whether slow evolution or rapid change, the fundamental economic factors in agricultural prosperity or adversity must remain the same. Prices are more than politics, and lowered costs than "land reform." It may be that a shrinkage in agricultural production over the world will lead to a relative rise in agricultural prices. If not, unless the value of money be stabilized and a downward drift of the general price level be prevented, agriculture must be depressed. In monetary policy the interests of farmer and manufacturer, of labourer and artizan, are identical.

Even with stable prices, it is not likely that, in an otherwise wealthy country like England, farming will ever be profitable enough to buy men of ability, character and courage. But our agriculture does not depend on economic motives alone. It can seldom offer riches, but it can promise a life full of natural interests, of healthy work lightened by country sport. The amenities that attract able men are a real asset, social and economic, to the rural community. Those who love the countryside live there, wholly or in part, for these less material rewards: for the beauty of the land and the quiet dignity of its ancient buildings; for the historic continuity that underlies the familiar routine of the farmer's year; the scent of new-mown hay in a midsummer meadow; red wheat ripening to harvest, and lines of living soil turned by the plough; autumn woods, glowing with bronze and gold and purple in the low November sun, as the first cock pheasant swings over the trees; the white coverlet of snow when, for a few short days, the land lies sleeping in the arms of winter; the flash of scarlet and the music of the hounds as they stream out of covert on a still, grey February day, while the earth waits for the discordant tuning of the March winds before breaking forth into the tumultuous and triumphant symphony of another spring.

Ignorance may injure England's pleasant land, but some of its joys will remain so long as men till the soil, and seed-time and harvest, and cold and heat, and summer and winter, and day and night do not cease.

# INDEX

9 781107 615526